BETWEEN EARTH AND SKY

The publisher gratefully acknowledges the generous contribution to this book provided by the General Endowment Fund of the University of California Press Foundation.

BETWEEN EARTH AND SKY

OUR INTIMATE CONNECTIONS TO TREES

NALINI M. NADKARNI

UNIVERSITY OF CALIFORNIA PRESS
BERKELEY LOS ANGELES LONDON

University of California Press, one of the most distinguished university presses in the United States, enriches lives around the world by advancing scholarship in the humanities, social sciences, and natural sciences. Its activities are supported by the UC Press Foundation and by philanthropic contributions from individuals and institutions. For more information, visit www.ucpress.edu.

University of California Press
Berkeley and Los Angeles, California

University of California Press, Ltd.
London, England

For acknowledgments of permission to reprint previously published poems, please see credits on page 299.

Library of Congress Cataloging-in-Publication Data

Nadkarni, Nalini.
Between earth and sky : our intimate connections to trees / Nalini M. Nadkarni.
p. cm.
Includes bibliographical references and index.
ISBN 978-0-520-24856-4 (cloth : alk. paper)
1. Trees—Social aspects. 2. Trees—Religious aspects. 3. Trees—Environmental aspects. I. Title.
QK477.N33 2008
582.16—dc22 2008002162

Manufactured in the United States of America

17 16 15 14 13 12 11 10 09 08
10 9 8 7 6 5 4 3 2 1

The paper used in this publication meets the minimum requirements of ANSI/NISO Z39.48–1992 (R 1997) (*Permanence of Paper*).

To the maples outside the front door, where this book began.
And to Craig Carlson, who urged me to move my desk closer to the window.

Trees are the earth's endless effort
To speak to the listening heaven.

—*Rabindranath Tagore,* Fireflies

I have a mind to confuse things,
unite them, bring them to birth,
mix them up, undress them,
until the light of the world
has the oneness of the ocean,
a generous, vast wholeness,
a crepitant fragrance.

—*Pablo Neruda, "Too Many Names"*

CONTENTS

ACKNOWLEDGMENTS

I thank Jade Leone Blackwater, whose help with research and editing was as valuable as her understanding and passion for this topic. I thank Jack Longino, my husband, and August and Erika, our children. Early encouragement came from James Fitzgerald. John McLain, Laurie Weisman, Susan Phillips, Ann Eriksson, Saroj Ghoting, Bill Yake, Larry Hamilton, and Maureen McConnell offered valuable editorial help along the way. Students in my "Trees and Humans" program and in the Forest Canopy Lab at the Evergreen State College provided information and ideas. A special thanks goes to Willy Fenske Towanda and Scott Hollis for help with images and permissions. Jono Miller, Claudia Mauro, Tim Scoones, Jack Ewel, and Anne McIntosh gave advice and support. John McLain offered valuable suggestions for the the title. Jenny Wapner, Dore Brown, and Anne Canright, from University of California Press, were incredibly helpful and skilled editors and greatly honed and improved the text. Funding was provided by a John Simon Guggenheim Foundation fellowship, the Helen R. Whiteley Center at the University of Washington, the National Geographic Society Committee on Research and Conservation, and a National Science Foundation Opportunities to Promote Understanding through Synthesis grant (DEB0542130). John Francis of the National Geographic Society Conservation Trust Program was an early believer in the approaches presented here. I thank the faculty, students, and staff of the Evergreen State College for providing the interdisciplinary habitat and support for the creation of this book. Finally, I thank my mother, Goldie H. Nadkarni, who encouraged me to climb tall trees and pursue my ideas.

Introduction

VIEW FROM THE TOP

> Let me stop to say a blessing for these woods . . .
> for the way sunlight laces with shadows
> through each branch and leaf of tree,
> for these paths that take me in,
> for these paths that lead me out.
>
> —*Michael S. Glaser, "A Blessing for the Woods"*

Strong brown hands clutched the blue climbing rope, knuckles white. Emil Arnalak, an Inuit born and raised in the tundra of Nunavut, in the far north of Canada, was holding on tight to his lifeline. Above him rose the treetops of a lush coastal forest in Washington State, while my students and I stood six feet below where he dangled. We were teaching Emil how to climb trees so that he could experience the forest canopy, the little-explored world high above our heads. But this was too strange, and he froze: "Too high! Too high! For me, too high!" he called in the singsong intonation of his native Inuktitut. Until the previous day, Emil had never even seen a tree, and he had rarely climbed anything higher than the stairs of the two-story buildings in his small village of Arviat. By merely being in the forest, he was exploring new territory. By climbing into the canopy, he was entering another world.

I had invited this tundra dweller, along with several other people unfamiliar with the life of a forest, to join me here for both my own professional expansion and personal curiosity. As an ecologist who has spent her life doing research in and teaching about forests, I wanted to learn how people who have never seen trees might perceive and value them. Would they find trees frightening? Boring? Useful? Beautiful? Would they be able to discern the variety and complexity of life forms in a forest? Would their categories of colors, textures, and shapes be the same as the ones I routinely use, or would they generate a different way of classifying the green around them—one with potentially different insights? The idea had come to me while I was climbing Mt. Rainier the previous summer. As I struggled up a snow-blanketed ridge of the tallest mountain in the Pacific Northwest, I diverted my mind from my aching legs by categorizing and giving my own names to the different types of snow I slogged through: snow that came in humps, snow that came in lumps, snow that stuck to my darn crampons. Of course, people familiar with wintry conditions have already created names for these and other different types of snow, and I wondered if my taxonomy would match theirs. Perhaps my status as a "snow novice" would provide them with new insights. That led to the idea of developing my own awareness and understanding of forests by listening to "forest novices."

So in the summer of 2003 I invited a group of Inuits, artists, musicians, rap singers, modern dancers, loggers, opera composers, and blind people—along with a cadre of forest ecologists—to observe the forest canopy anew. Our staging area was Ellsworth Creek, a reserve of primary and secondary forests in southwestern Washington owned by the Nature Conservancy. My students and I rigged trees with climbing ropes and hoisted up four plywood platforms for participants to sit on. On the ground, we set up tents, a cooking shelter, and a campfire pit and prepared for two weeks of living in the forest. The National Geographic Society supported this endeavor through the Conservation Trust, a program that invests in projects that might lead to new approaches to conservation.

Emil and his fellow Inuit, Brian, had traveled for four days from their village to reach our site. During our drive from the airport to Ellsworth Creek, I learned that their native language of Inuktitut had—as expected—

over twenty-five words for snow, including all of the snow types I had encountered on my ascent of Mt. Rainier. However, they had no word at all for tree. "We use the word *nabaaqtut*," Emil explained, "which means 'pole.'" And forest? "We use *nabaaqtut juit*, which means 'many poles.'"

Now, my students and I waited for Emil to continue upward. I paralleled his climb on a companion rope, encouraging him to trust the gear and to trust the tree. Slowly, we made our way to the platform hanging sixty feet above the forest floor. Emil pulled himself over the handrail and settled himself on a corner of the platform, taking the time he needed to regain a feeling of ease. He watched the swaying curtain of foliage that enveloped the two of us. He observed the sun reflected in the drops of dew on the underside of the spruce needles, and pointed out the pattern of sword-fern leaves arrayed, fanlike, below us. After an hour he nodded twice, then asked to descend. In the days that followed, Emil never climbed again, but I saw him walk up to certain trees—most frequently, western red cedars, the species most valued by the Native Americans of my own region—and place a hand on them for minutes at a time. He spoke with each of the project participants about the forest and took notes in the hieroglyphic-looking script of Inuktitut in the notebook we had given him.

At the end of our time in the forest, we lit a last celebratory campfire. Each participant shared a song, a plant collection, a drawing, a poem, or a bit of scientific knowledge. When his turn came, Emil slowly stood to his full height of just over five feet. He looked at each of us in turn and, in a formal tone, said, "In these days, I have learned that trees are more than just poles. You must learn to treat these big trees the way we treat the elders in our village—with great care and great respect. Trees are as important to you as our grandparents are to us because they teach you things." A long pause followed Emil's words, given weight by the strength of its source—a man who until the week before lacked a word for *tree* or *forest.*

This book expands on Emil's statement about the powerful connections between humans and trees, exploring the many ways humans use trees and the lessons they teach us. It has grown from my years of teaching and conducting scientific research on the ecology of rainforests. But just as importantly, I wrote it because I love trees: how they look, how they behave, how they smell and sound, and how I feel when I am around them. When I place my own strong brown hand on the trunk of a tree, I feel connected to something that deserves my curiosity, care, and protection. It is my hope that this book might do for readers what those cedar trees did for Emil: awaken—or reawaken—a sense of wonder and respect.

BEGINNINGS

O profound
Silent tree, by restraining valor
With patience, you revealed creative
Power in its peaceful form. Thus we come
To your shade to learn the art of peace,
To hear the word of silence; weighed down
With anxiety, we come to rest
In your tranquil blue-green shade, to take
Into our souls life rich, life ever
Juvenescent, life true to earth, life
Omni-victorious.

—*Rabindranath Tagore, "In Praise of Trees"*

My journey to understand trees started early, in the front yard of my childhood home in suburban Maryland. Most afternoons after school, I exchanged a hellohowwasschoolfine with my mom, then grabbed a snack and a book to read. Back outside, I would choose one of the eight maple trees that lined the driveway to climb, each with its own vertical pathway to a comfortable nest aloft. My favorite, which I had named "BigArms," had a ladderlike set of branches nearly to the top, whereas "Skinnyhead" had a problematic crossing I was incapable of negotiating until I reached a critical height, climbing with a speed and confidence that drove my severely acrophobic mother into retreat. Those perches aloft were my refuge from the world of homework, parental directives, and the ground-bound humdrum of the everyday. I could look out across my home territory, check on the progress of squirrel nest constructions, and feel the strong limbs of those trees holding me up for as long as I wished. When I was nine years old, I wrote, illustrated, and published a single copy of a book titled *Be Among the Birds: My Guide to Climbing Trees.* In retrospect, I understand that my sense of caring and kinship with trees germinated during those afternoons of arboreal repose.

Trees were not my only focus in those years. I took modern dance lessons from Erika Thimey, a German-born teacher who offered the gift of creativity to her students. I learned the expressive ways the body can move and acquired the discipline I needed to hone my muscles. From Miss Erika, as we called her, I learned that, with mindfulness, the simple act of walking across a wooden floor or noting the grace in a falling leaf can be an aesthetic action. One of our assignments was to make a dance about negative space—the space between objects. Seeing the world not as solids but as shapes of air opened up a whole new perspective on things, one that has kept me aware of the multiple ways that one must look at nature to understand it fully—an approach I brought to the writing of this book.

In college, I first encountered the world of forest ecology in the lectures of a behavioral ecologist, Dr. Jon Waage. When he wasn't teaching Brown University undergraduates, he carried out research on damselfly behavior. I was amazed to learn that he could make a living by sitting at stream edges recording the movements of aquatic insects. From him, I learned about academic science. He posed seemingly narrow questions that later turned out to relate to much broader issues of life and death, competition and mutualism, and evolution. How does a female's wing size affect mate choice? How does mate choice affect population diversity? How does diversity affect resilience in the face of disturbance from human activities? Traversing the labyrinth of the scientific literature, I learned to trace citations to their sources and recognize the key players in a scientific discussion. Attending professional meetings gave me a tribal

1. Six-year-old Nalini Nadkarni peers from a tree. Photo by M. V. Nadkarni.

sense of community as I listened to the elders and became initiated into the rituals and hierarchies of the discipline.

I loved the challenge of untangling the endless puzzles I encountered in nature. Like veterinarians with dogs and obstetricians with the soon-to-be-born, ecologists explore their subjects without the benefit of mutually intelligible speech. They rely instead on their own observations and the written observations of others, bound in books and archived in libraries. The new Sciences Library at Brown University was a tall, imposing structure that I sought to make familiar. Early in my senior year, I spent a night there, hiding out in the bathroom with my sleeping bag until the doors were locked and the custodians departed. Starting on the top floor, I walked slowly through the stacks, Astronomy to Zoology, touching the book spines, dipping into a tome here, a journal there. As dawn broke, I found myself on the ground floor, tired but happy to have made at least a fleeting acquaintance with the realm of science as represented in its printed form. What I absorbed that night confirmed what I learned from Dr. Waage. Science—a word that comes from the Latin *scire,* to know to the fullest extent possible—is one good way to understand and communicate what interests me about the world. And so I joined the world of science, not to gaze at damselflies like Dr. Waage, but to study trees and forests.

After graduation, I tried on the life of a field biologist, accepting a position, for $75 a month, as assistant to a septuagenarian entomologist named J. L. Gressitt, who studied the taxonomy of tropical leaf-feeding beetles and directed the Wau Ecology Institute, a tiny field station in the highlands of Papua New Guinea. I earned money for my airplane ticket across the Pacific by working for four months in logging camps in southeastern Alaska doing preconstruction surveying, flying out in floatplanes to tiny settlements nestled in the temperate rainforest near Ketchikan.

In October 1976, I traveled to the foothills of Morobe Province. Dr. Gressitt's field station consisted of a few shabby wooden buildings, a small herbarium and insect collection, and a central table occupied by a chipped coffee pot around which staff gathered each morning to discuss their research projects. I spent the next twelve months accompanying Dr. Gressitt on his expeditions around the country, he with a pair of forceps to pluck leaf-feeding beetles out of the foliage, I in his wake with a plant press to collect the host plants. I filled my spare time learning all I could from the other researchers and our tiny library. Several times, I felt the jolt that follows a small but real discovery about nature, a steppingstone of understanding.

My passion for dance was still very much alive, however, and after the year in that rainforest cloister, I went to Paris, where, soon, I received an invitation to practice with a modern dance company, Danse Paris. It was pure delight to gulp down the cultural offerings that only Paris provides. I gazed at gems of urban architecture, masterpieces of art in museums, and the wares on display in bakery and shoe store windows. I marveled at the diversity of human-made objects, just as a few months before I had stared in equal amazement at the diversity of orchids and ferns that draped the branches of tropical trees in Papua New Guinea.

After six months, I had to make a choice: the forest or the stage? As much as I loved the world of dance, the time I spent in the tropical rainforests seemed truer to my spirit. I also felt closer to my biologist colleagues than to my fellow dancers, and more at peace in the forest environment than in an urban milieu.

In 1979, I returned to the United States and enrolled as a doctoral student in the University of Washington's College of Forest Resources. During my first summer, I took a field course in the tropical biology of Costa Rica. Our class visited forest sites with biologists who ran field projects. Whenever we entered a rainforest trail, my eyes were drawn upward to the plants and animals in the treetops, provocatively out of reach. A score of questions burbled up. How did the plants live up there without connection to the soil? What exactly were the birds feeding on? Who was pollinating which flowers? Were there insects that lived their whole lives up there? How did the different species of monkeys interact in the treetops? My instructors had no answers to these questions, because almost no one had studied—or even climbed into—the tropical forest canopy. Most of these trees have unnervingly tall trunks, without lower branches, and can sport spines, biting insects, and the occasional lurking snake. The tree-climbing skills I had developed in the benevolent trees of my childhood were useless.

At one of the field stations we visited, however, I encountered a graduate student, Don Perry, who *was* studying the canopy of a lowland forest. He had applied mountain-climbing techniques to gain safe and nondestructive access to the canopy, using an approach that had been pioneered just a few years before by ecologists in the Pacific Northwest. As I watched Don climb, seemingly without effort, from the ground to a small wooden platform 125 feet above, I knew that that was what I wanted to do too: climb trees and explore the forest canopy. My aspirations were timely; Don needed a field assistant, and he offered to teach me how to climb in exchange for a month of help.

Two observations stimulated my mind and spirit the first time I found myself perched in the crown of a huge *Virola* tree looking out over the canopy below me. The first was the sheer abundance and diversity of canopy-dwelling plants—orchids, bromeliads, ferns, mosses, lichens—

2. *Tall tropical trunks lack lower branches but foster a rich diversity of canopy plants, or epiphytes. Photo by author.*

which covered every available stem and branch like so much floral upholstery. These plants, called epiphytes, have no connection to the vascular system of the tree (the living tubes and chambers responsible for transporting water, sugars, and minerals from one tree part to another). Rather, they derive water and nutrients from rainfall that is intercepted by their foliage. Second, even though I knew that these plants are independent of their host trees, I sensed that they contributed to the forest ecosystem in ways that had not been documented by ground-bound forest ecologists. I wanted to find out how these arboreal communities might function as independent but interacting subsystems of the forest.

When I returned to Seattle, I pursued a dissertation project that involved a comparative study of the temperate rainforest canopy of Olympic National Park and the tropical cloud forest canopy of Costa Rica. Cloud forests and temperate rainforests both support extensive communities of epiphytes, though the types and species of plants found in each kind of forest are very different. For four years, I collected epiphyte samples and analyzed their nutrient content to quantify the biomass and nutrient capital held within the epiphytes relative to the whole ecosystem.

This was a first step in coming to understand the role of epiphytes in forest ecosystems, and over the past three decades I have worked to expand our knowledge of how forest canopies function. New techniques of canopy access have evolved to include hot-air balloons, treetop walkways, hanging platforms, and thirty-story construction cranes. My colleagues, students, and I have enumerated the rare species that, dwelling as they do high on branches and twigs, never appear in ground surveys. We have learned that treetop versions of traditionally terrestrial invertebrates—beetles, ants, springtails, and even earthworms—are found in the arboreal soil that accumulates on tree limbs beneath mats of epiphytes. Canopy studies have revealed that birds of the cloud forest use epiphytic flowers and fruits for over one-third of all of their foraging visits, underscoring the importance of these plants in the food web.

Such research confirms that trees are a critical part of ecosystems, landscapes, and the biosphere. Canopy biologists now quantify the amount of oxygen tree canopies produce, the amount of carbon dioxide they store, the volumes of soil they protect, the amount of water they retain, and the scores of wildlife species they support. Urban foresters have documented the "ecosystem services" provided by tree canopies in urban settings: reduction in noise, temperature, and pollutants. The pivotal role played by trees ensures that loss of canopy diversity and function is a loss both to forests and to the landscapes beyond.

THE HIERARCHY OF NEEDS

I am looking at trees
they may be one of the things I will miss
most from the earth
though many of the ones I have seen
already I cannot remember

> and though I seldom embrace the ones I see
> and have never been able to speak
> with one
> I listen to them tenderly
> their names have never touched them
> they have stood round my sleep
> and when it was forbidden to climb them
> they have carried me in their branches.
>
> —*W. S. Merwin, "Trees"*

We have a clear affinity—the word comes from the Latin *affinis*, indicating relation by marriage—for trees. Although we are not of the same family, trees and humans are in a sense married into each other's families, with all the challenges, responsibilities, and benefits that come from being so linked. From the first glimmers of humanity's dawn, we have evolved with trees. Our grasping hands, opposable thumbs, and binocular vision once enabled us to leap between branches with easy confidence, though now we use these endowments for other things, such as playing baseball and dodging in and out of traffic. The sky-high cost of penthouse apartments reflects our desire to look out and over the landscape—a desire born in the tree-studded savannas of our evolutionary cradle.

In this book, I explore these affinities. In 2003, I taught an interdisciplinary class called "Trees and Humans." One day I invited Bill Yake, a tall, gentle-voiced naturalist and poet, to present a guest lecture on how trees are portrayed in poetry. He began by talking not about trees or poetry but about a psychologist, Abraham Maslow. In 1943, Maslow wrote a paper titled "A Theory of Human Motivation," which soon became a classic in the field. Earlier researchers who sought to explain what inspires humans to act had focused on factors such as genetics or particular achievement standards such as earning a doctoral degree or reaching a high income level. Maslow, in contrast, posited that humans aspire to "self-actualization," which involves a focus on solving problems, nurturing an appreciation of life, achieving personal growth, and seeking out peak experiences. Maslow drew up a hierarchy of human needs in the shape of a pyramid composed of five levels. The broad base of the pyramid consists of physical needs (food, water); next comes safety (shelter, stability); then social needs (the desire to be included, loved); then "ego" needs (self-esteem, recognition); and finally, at the apex, self-actualization (personal growth, creativity). For self-actualization to be realized, each previous lower-level need must first be satisfied.

In his lecture, Bill identified the human needs that trees fulfill at each level of Maslow's hierarchy, as reflected in poetry about trees. For example, Li-Young Lee's poem "From Blossoms" illustrates how trees fill not only our physical needs, by providing us with fruit, but also our social needs, by bringing humans together in an orchard, at a fruit stand, over a shared meal:

> From blossoms comes this
> brown paper bag of peaches
> we bought from the boy
> at the bend in the road where we turned towards
> signs painted *Peaches.*
>
> From laden boughs, from hands,
> from sweet fellowship in the bins,
> comes nectar at the roadside, succulent
> peaches we devour, dusty skin and all,
> comes the familiar dust of summer, dust we eat.

After the class was over, I reflected on Maslow's hierarchy and Bill's application of it to poetry. The five levels of human needs struck a chord in me, because I could see so clearly how trees satisfy us on every level. Over the ensuing months, I began to modify the original hierarchy to reflect my own understanding of how human needs are met, in particular by trees. This is evident in many forms of human expression: scientific knowledge, art, literature. My version of the hierarchy incorporates a few levels that Maslow did not include, such as play, imagination, and spirituality. I have also reconceived Maslow's apex of "self-actualization"

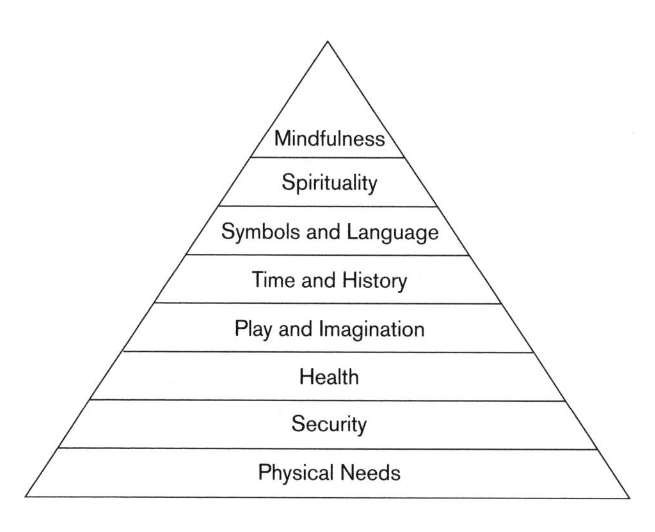

3. The structure of this book is based on a modification of Abraham Maslow's hierarchy of needs.

as a state of "mindfulness," which I define as the need to be aware of and compassionate toward one's surroundings. I have organized the chapters of this book around this hierarchy.

As in Maslow's model, in my hierarchy the first levels of human need fulfilled by trees are physical. In chapter 2, I describe the ways that trees provide us with air, food, water, and other material necessities. The oxygen released from the stomata of leaves fuels every process in our bodies. Trees make the oranges for our breakfast juice and the apples we pack in our lunches. Tree products—the brown bags that contain those lunches, the cartons for our milk and corks for our wine bottles—help us preserve and transport food and drink. Trees are indirectly responsible for transporting information, as well. The classified ads of the newspaper that my son scanned for summer jobs this morning are printed on paper derived from the pulp from conifer trees harvested in southern Georgia.

In chapter 3, I discuss other sources of physical security—shelters, fire, windbreaks, and a sense of place—that come from trees. Overarching branches, for example, may provide a bower for sweethearts, a secure and protective venue in which they can express their emotions: "Love, meet me in the green glen," wrote the poet John Clare, "Beside the tall elm tree, / Where the sweetbriar smells so sweet agen; / There come with me." More practically, we build our homes and offices with lumber, using studs excised from the trunks of mighty Douglas-firs and plywood rolled off of the slimmer trunks of southern pines. Wood from trees fuels fire, providing warmth and light that help humans maintain their social health in the cold dark times of winter. The crackle of a woodstove on a wet February night causes my family to gather in our living room, each of us grateful for the firewood that my husband has felled and my son has split. Trees also provide a sense of place, of rootedness. The long lifespans of trees and their prominence in the landscape make them significant landmarks for humans.

The next levels of need concern our health, as discussed in chapters 4 and 5. In chapter 4, I show how trees provide chemicals used by indigenous people around the world for healing, and by practitioners of Western medicine as well. Researchers have found that tree imagery reduces the stress experienced by hospital patients. Another level of human needs that grown-ups too often forget is the need for fun and for a healthy imaginative life, which is the focus of chapter 5. By the time we enter our twenties, most of us have put aside our toys and become observers rather than participants in games and sports. However, trees provide spaces and tools for lifelong play, for both recreation and the re-creation of ourselves. A tree house can give a child a critical sense of freedom. Even in the most urban of cityscapes, humans have set aside land where children and adults can enjoy reflective walks, energizing jogs, community gatherings, and "just playing." Central Park, for example, is an invaluable island of nature in the middle of New York City, as important to and symbolic of that city as the tallest skyscraper. In his essay "On Trees," the German poet Hermann Hesse wrote: "When we have learned how to listen to trees, then the brevity and the quickness and the childlike hastiness of our thoughts achieve an incomparable joy. Whoever has learned how to listen to trees no longer wants to be a tree. He wants to be nothing except what he is. That is home. That is happiness."

Moving up the hierarchy, we come to our need to place ourselves in time and to use language and symbols. In chapter 6, I discuss the many ways trees help people cultivate a sense of history. The long lives of trees give us a sense of continuity that traverses generations, and the grand scale of time for trees and forests can help us put our own actions and ambitions in perspective. In chapter 7, I explore representations of trees in language, art, and music. The word *tree* emerged from the Sanskrit word *dāru*, which means both "wood" and "to be firm or solid." Our idiomatic expressions reflect the association we have made between trees and truth—we *get at the root of things* and *avoid barking up the wrong tree.* Peoples of ancient times recognized the symbolic power of trees, and they crafted magical things of wood: wands, staffs, scepters, and divining rods. Today we still use tree imagery to adorn items of value; trees or parts of trees can be found on the postage stamps and currency of over sixty countries.

Many humans have a sense of spirituality, an awareness that we are linked to something larger than ourselves. Trees play a role in the spiritual life of many cultures, as I explore in chapter 8. The *axis mundi*—the central pivot of the cosmos, or imaginary line linking heaven and earth—takes the form of a tree in a number of cosmologies. The Tree of Life and the Tree of the Knowledge of Good and Evil are introduced in the very first book of the Bible; these and other mythological trees confer wisdom or immortality on humans. In some countries, the only remaining trees of certain species are rooted in sacred groves or churchyards, where they were protected because they symbolized eternal life, provided a conduit for communication with the gods, or were deities themselves. The Buddha achieved enlightenment as he sat under the spreading limbs of the Bodhi tree, breathing in and out in silence, as does a tree.

As we move into the new millennium, I—and many other scientists, educators, and artists—observe with concern the growing distance between humans and the natural world, especially forests. My own teenage son, whose first tooth emerged during a family backpacking trip in the Cascade Mountains, chooses not to emulate my childhood pastime of climbing trees after school. Instead, he logs on to the computer for a virtual chat with friends, iPod to ear, cell phone charging in its plastic nest beside him. More and more of the interactions we do have with nature are mediated by technology. This increasing distance between nature and humans is both a cause and a consequence of the severe environmental problems we carry forward into the next century. I hope that recognition of the human affinity for trees—and by extension, other parts of nature—may help decrease that distance. This hope explains why I have placed mindfulness at the apex of my pyramid of human motivation, and why it is the final discussion in chapter 9. As people become increasingly aware of trees' remarkable biology, the diverse array of goods and services they provide, the complexity of meanings they symbolize, and the art they have inspired, perhaps we will at last heed Emil's quiet campfire request to respect our old trees.

Craig Carlson, a poet and colleague of mine, introduced himself to me with the following message:

> . . . there is a good poem about trees that go whinneying like horses in the wind by I think William Carlos Williams, which is exactly right. We live near an old growth grandfather fir right next to where we sleep that gallops through the night anytime a breeze kicks up amazing grace it is the sound of spiritual horses, like angels or gods . . .

Craig and others have helped me appreciate poetry as an articulation of that which is holy or beautiful or terrible to humans. Science has been a domineering force in my life, but lately I have decided that I want to do more than simply conduct experiments, procure grants, write papers, and test hypotheses to the 0.05 level of significance. Craig told me to relax, to stretch, to listen to the other voice in me, to ask for a desk with a view.

4. The mist of the cloud forest in Braulio Carrillo National Park, Costa Rica, provides a serene setting for science and contemplation. Photo by Greg and Mary Beth Dimijian.

And so I have included poems in this book, so that poetry joins science in helping us understand and appreciate trees.

> Green I love you green. Green wind. Green branches.
>
> —*Federico García Lorca, "Romance Sonambula"*

Spirituality, inspiration, affinity—these words are not regularly bandied about at the scientific symposia and academic meetings I attend to share results of my forest canopy studies. Many of my colleagues might even lean over and whisper that it would be considered an academic faux pas to admit to a personal affinity for one's object of study. Science, after all, is about being objective, about looking at the hard facts. But my own way of knowing the world has included much more than just my professional life. It has embraced dance and family, wintertime trail running and sitting in front of a woodstove with a book of poetry. I offer this book as an invitation for you to consider how the approaches of science, art, and the humanities can, all together, move us toward mindfulness about trees.

Breath

Tree, gather up my thoughts
Like the clouds in your branches.
Draw up my soul
Like the waters in your roots.
In the arteries of your trunk
Bring me together.
Through your leaves
Breathe out the sky.

—J. Daniel Beaudry

Chapter One

WHAT IS A TREE?

> A tree says: A kernel is hidden in me, a spark, a thought, I am life from eternal life, unique in the smallest play of leaves in my branches and the smallest scar on my bark. I was made to form and reveal the eternal in my smallest special detail.
>
> —*Hermann Hesse, "On Trees"*

A person wishing to describe a tree and the environment around it has a deceptively difficult charge. *Webster's* dictionary defines *tree* simply as "a woody perennial plant having a single usually elongate main stem generally with few or no branches on its lower part." Most of us would point to other basic parts of a tree as being somehow characteristic as well: leaves, bark, roots. Indeed, trees can be as familiar to us as our own bodies. Trees and people are even built on the same general pattern: upright in form, with a crown on top and mobile limbs stemming from a central trunk. But this very familiarity can make it hard to define just what a tree is. Perhaps the difficulty arises because we take trees for granted, as simply part of us and our world. Or maybe we haven't developed the particular vocabulary needed to describe the complex joinings and curvings of branches, the clustering and nodding of foliage, and the nuances of the nearly infinite shades of green and brown that can change in an instant of breeze or dappling sun.

TREES FROM TOP TO BOTTOM

That one born in the forest, growing up
With canopies, must seek to secure coverings
For all of his theories. He blesses trees
And boulders, the solid and barely altered.
He is biased in terms of stable growth vertically.
And doesn't he picture his thoughts springing
From moss and decay, from the white sponge
Of fungus and porous toadstools blending?
He is shaped by the fecund and the damp,
His fertile identifications with humus
And the aroma of rain on the deepening
Forest floor. Seeing the sky only in pieces
Of light, his widest definition must be modeled
After the clearing hemmed in by trees.

—Pattiann Rogers, "The Determinations of the Scene"

My knowledge of the shapes and forms of trees is incorporated into the muscles and skinned knees of my youth, growing from my schoolgirl afternoons climbing maple trees in our front yard. Those early lessons in how trees are assembled complement my later formal studies in forest ecology. Tree structures can be seen not only in botanical trees, but in many other objects in nature and society. Sometimes magnified, some-

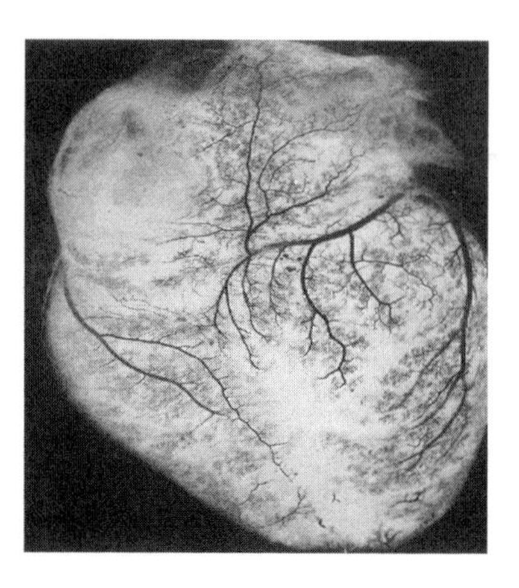

5. Dendritic patterns of veins and arteries, the vascular system of human bodies, are decidedly treelike. Photo courtesy Science Photo Library.

times miniaturized, tree forms are everywhere around us and within us. They exist in the mighty rivers of our landscapes, in the plumbing beneath our sinks, and in the passageways that lead us deep into caves. The walls of medical examining rooms are hung with images of the vascular system—trees writ large—allowing doctors to trace the flow of blood, lymph, and air through passages that mimic my maples. The family tree, the mathematical tree, and corporate hierarchies are all abstract forms that humans use to understand complex relationships. Tree forms even shape the way we think. Neurons are called "dendrites," from *dendron*, the Greek word for tree, and they convert outside stimuli into our awareness, our thoughts, and our memories. But before we explore these more abstract and human-centered realms, I invite you to look at the complex structures of real trees and real forests.

THE ABOVEGROUND WORLD

If I strapped you into a tree-climbing harness and taught you some basic climbing and safety moves, we could take a voyage from the forest floor to the top of my favorite tropical fig tree, "Figuerola," in Monteverde, Costa Rica. There we would view up close and personal the remarkable body parts of an individual tree and experience the tremendous variability of the living conditions it creates.

A mature rainforest tree such as Figuerola is composed of a set of complex structures, from the networks of root tips that penetrate the earth for six stories below it to the skinny branch tips thirty stories above the surface of the forest floor. These twigs support the foliage of the tree, which carries out the critical job of collecting and storing energy. Most trees present their leaves at the outer edges of their crowns. In simplest terms, these leaves are the energy factories of the tree. They contain light-capturing pigments—mainly chlorophyll—that absorb sunlight energy and carbon dioxide from the air, mix them with water absorbed from the soil, and convert them to sugars and oxygen. The sugar is used or stored in the branches, trunk, and roots. The oxygen is released into the atmosphere, where it is used by humans and other animals for respiration (the taking in of oxygen and exhalation of carbon dioxide). Trees also respire, but the carbon dioxide they create in the daytime is masked by their simultaneous production of oxygen via photosynthesis. At night, however, they produce carbon dioxide, just as do people, dogs, and microbes.

This complex process of photosynthesis has been of interest to botanists for nearly four centuries. As early as the 1640s, the investigations of both the Flemish chemist and physician Johannes Baptista van Helmont and the English clergyman and physiologist Stephen Hales showed that plants require air and water to grow. In the 1700s Joseph Priestly, another theologian and natural philosopher, carried out his famous experiment of putting a mouse and a sprig of mint under a bell jar. The mouse survived, which demonstrated that green plants can replenish oxygen-poor air, thus supporting respiration and life. Following up on this work, the Dutch physiologist and botanist Jan Ingenhousz learned that plants can create oxygen only in the presence of sunlight, indicating that it is not the heat of the sun but rather its light that fuels the process. Research in the nineteenth century produced information on the specific processes and materials involved, and in the twentieth century the complex biochemistry of photosynthesis was unraveled.

Most trees arrange their leaves to maximize the capture of sunlight, but their basic structure dictates how they do it. Certain trees, such as Figuerola, have their leaves distributed in a thin "monolayer" along the entire outer envelope of the crown, which creates a shady area just beneath them. Other trees, such as the evergreen Douglas-firs of my home forests, with their tall, narrow crowns, also create foliage that is located in the self-shaded portion of their crowns. Their needles can carry out photosynthesis with less than 10 percent of the sunlight that arrives at the outer edge of the canopy, allowing them to maximize the capture of light that filters into the forest.

The major structural support for all of this energy-gathering apparatus is the trunk. Like our own trunk, it forms the main portion of the tree, rising from the ground and then differentiating into the crown, giving the tree its overall shape and strength. It is the principal conduit for fluids that transport nutrients, hormones, and sugars up and down and around the tree, just as our own vascular system distributes nutrients and energy throughout our body. In a tree, this transport is accomplished not by arteries and veins, but via xylem and phloem, a network of tubes that run between the roots and the leaves, respectively carrying water and minerals upward and moving sugar down from the leaves to the branches, trunk, and roots for storage.

The trunk consists of a series of nested layers—in the form of concentric cylinders, with heartwood at the center and bark at the surface. As the tree grows, xylem cells in the central portion of the tree become inactive and die, forming the tree's heartwood. This material is chock-full of stored sugars and oils, and so it is usually darker than the younger sapwood that surrounds it. In species such as western red cedar, the heartwood is also packed with tannins, which make it extremely resistant to decay. The heartwood of bigleaf maple, in contrast, is soft and unprotected by resistant chemicals, so nearly every individual over sixty years old has a hollow core.

The sapwood, also made up of xylem, comprises the younger layers of wood. Its network of thick-walled cells brings water and nutrients up from the roots through a connected set of hollow cells within the trunk to the leaves and other parts of the tree. Immediately beneath the outer bark is the phloem, or inner bark, which acts as a food supply line by carrying sap (sugar and nutrients dissolved in water) from the leaves to the rest of the tree. The all-important cambium is a very thin layer of tissue between the xylem and phloem; it produces new cells that become either xylem, phloem, or more cambium. The cambium is what allows the trunk, branches, and roots to grow larger in diameter.

A tree's outer bark is the functional equivalent of our own skin. It originates from phloem cells that have worn out, died, and been shed outward, and it acts as a first line of defense against insects, disease, storms, and extreme temperatures. In certain species, the outer bark also protects the tree from fire. This remarkably variable material can differ greatly in thickness, texture, and color, from the thin, papery bark of birch to the thick, multilayered bark of Douglas-fir, which can withstand fire-generated temperatures up to 1200°F.

The limbs that attach to the trunk provide a network of woody material that in turn supports the leaves. Typically, branches are formed by the sprouting and extension of terminal and lateral buds at the tips of twigs. In tree species that grow in northern temperate regions, such as maples, these buds are created in the late summer and lengthen in the following spring. Tropical trees, whose annual rhythms are more subtle because of the lack of pronounced seasonality of temperature, are less predictable in the timing of bud burst. Some trees, such as Douglas-fir in the Pacific Northwest, are capable of sprouting "epicormic branches." These emerge from latent buds or from the cambium itself after the trunk has been exposed to extra sunlight following damage to its branches by a falling neighbor. Epicormics tend to have a very different structure than normal branches, often growing out in a fanlike and stubby shape that provides a horizontal structure on which birds may make their nests. The ability of Douglas-fir to form epicormic branches may explain these trees' amazing longevity—1,200 years is not an unusual age for a Douglas-fir in the absence of fire, severe windstorm, or the buzz of a chainsaw.

ARBOREAL HITCHHIKERS: THE EPIPHYTES

Branches and trunks also provide a home for epiphytes, "plants that grow on plants." I think of them as tree hitchhikers in slow motion—very slow motion—as they take up space but neither provide energy, nutrients, or water directly to the vascular system of their accommodating hosts nor take them. In contrast, true parasites such as mistletoes have specialized roots called haustoria that penetrate the bark and outer stem to draw water and energy from the host. Although epiphytes are sometimes called "air plants," they don't really live on air. Rather, through their leaves, they are able to take up dissolved nutrients that, delivered in rainfall and mist, occur in dilute concentrations. Some of their nutrient needs are served by rain that falls through the crown and leaches nutrients from intercepted host tree leaves that have fallen and decomposed within the interstices of the tree.

The distribution of epiphytes often reflects the microclimate gradients that exist along branches. A tree's microclimate—a set of specific environmental conditions—varies greatly with its height. At the base of a tree, the shade cast by the tree itself and by surrounding vegetation creates a microclimate that is damp, dark, still, cool, and moist. Growing on the moist lower trunk we might find delicate-leaved orchids in the shade-loving genus *Gongora*, which can tolerate almost no stress from sunlight-induced drought. Higher up in the tree, temperature, wind, and sunlight increase in intensity, and we find epiphytes such as *Peperomia*, which has thick succulent leaves that resist water loss. One of the intriguing insights I had when I first began climbing is that the same pattern of increasing temperature, wind, and sun intensity occurs as one moves out from the trunk onto the branches. Perched on branches at the outer envelope of the canopy we might encounter individuals in the cactus family—the genus *Epiphyllum*—looking for all the world as though they were resting stolidly on the sand of an Arizona desert.

And then there are plants that, Zelig-like, completely change their morphology in response to different microenvironments. Vines in the genus *Monstera*, for example, begin their lives on the cool, dark forest floor. Leaves of the juvenile plant grow in an alternating pattern, pressing themselves completely against the supporting trunk, creating a ladder by which the vine grows up to the sunny, windy conditions of the canopy. There, they produce thick, floppy leaves the size of a large pizza, with deep, irregular indentations along the sides and large blob-shaped holes in the center, a shape that reduces damage from the high winds the plant must endure in the canopy.

Recent experimental research by Martin Freiberg, a curious and energetic German ecologist at the University of Ulm, revealed that epiphytes can themselves affect the microenvironment within the crown, even at the spatial scale of a hummingbird's wing. He draped recording equipment on pairs of branches—half of which he stripped of epiphytes, half of which he left intact—and collected measurements of temperature and humidity over four weeks. He found that air temperatures around stripped branches became hotter by as much as 8°F during the day, and up to 5°F at night. Why? The mechanism Martin proposed is that the densely packed epiphytes reduced air circulation, trapping unheated air on shaded branch undersides instead of allowing it to mix around the branch.

Epiphytes were shown to have a cooling effect in another situation as well. A team of German and Swiss researchers recently documented what they called "rainforest air-conditioning," a moderating effect of epiphytes on branch microclimate. Sabine Stuntz, Ulrich Simon, and Gerhard Zotz measured the temperature of the branch surface and the drying rate of leaves at various locations within tree crowns that either had different epiphyte assemblages or were epiphyte-free. They found that during the hottest and driest time of day, sites next to epiphytes had significantly lower temperatures than locations within the same tree crown that were bare of epiphytes, even though the latter were also shaded by host tree foliage or branches. Moreover, water loss through evaporative drying at sites next to epiphytes was almost 20 percent lower than at exposed microsites. Although the influence of epiphytes on temperature extremes and evaporation rates is relatively subtle, their mitigating effect could be of importance for small animals, such as the insects that inhabit an environment as harsh and extreme as the tropical forest canopy.

It turns out that epiphytes have a disproportionately large influence on the whole forest, relative to their small size. In the late 1970s, research into the ecological roles of epiphytes in the nutrient cycles of the entire forest—pioneered by Bill Denison and George Carroll of Oregon State University and the University of Oregon, respectively—looked at old-growth forest canopies in the Pacific Northwest, focusing on the relationships between forest structure and function. They determined that canopy-dwelling plants and lichens augment the amount of nutrients captured in the atmosphere, by "fixing" gaseous nitrogen and transforming it into forms of nitrogen—nitrate and ammonium—that can be used by plants and animals to build and replace their body parts. In Monteverde, Costa Rica, my students and I have measured the amounts—sometimes considerable—of nitrogen and other nutrients that the epiphytes intercept and retain from rain, mist, and dust. One study, for example, documented that 60 percent of all the nitrate that enters the ecosystem in rain and mist is sucked up and held in the canopy by branch-dwelling mosses. Thus, although epiphytes may be a barely visible and seemingly negligible component of forests to ground-bound scientists, in fact they perform a keystone function.

Last but not least, our inventory of tree structural elements includes arboreal soil. We expect to find soil on the forest floor, where tree roots are embedded. But in many wet forests, such as Monteverde, where Figuerola grows, we encounter pecks and even bushels of soil on the large branches that support epiphytes. In a temperate rainforest, within the single crown of a mature maple tree, this soil can amount to as much as 280 pounds (dry weight). Where does this material come from? When the epiphytes die and decompose, they generate a layer of soil up to ten inches thick that rests on canopy branches. This soil provides a habitat for a huge diversity of insects, earthworms, and spiders, which in turn are critical sources of food for birds and tree-dwelling mammals.

THE FOREST UNDERWORLD

Roots and their associated animals and fungi are the most poorly understood portion of forests, overlain as they are by opaque soil that obscures the fascinating interactions going on beneath. Roots take many forms, from massive buttresses as imposing as a three-story cliff face to tiny root hairs just eight one-thousandths of an inch in diameter. Soil inhabitants—some, such as nematodes, mites, and bacteria, smaller than pinpoints, others as big as badgers—participate in food chains and energy cycles as complex as the subterranean networks of electrical cables that operate beneath the streets of great cities. Just as canopy researchers have learned how to safely climb into the tops of trees, scientists have recently invented techniques for understanding how to measure and interpret the belowground world.

One weekend in 1980 I accompanied a fellow graduate student, Mike Keyes, on one of his collecting trips into the field. He was studying the dynamics of fine roots (those less than a quarter of an inch long, and responsible for nearly all of the water and nutrient absorption of trees) in an alpine forest in Washington State. He had constructed a "rhizotron"—literally, a place to observe roots grow. Descending a ladder into the dark room that Mike had dug out twenty feet belowground, I felt my claustrophobia kick in with a vengeance. My request to bring up the lights received a quick "nope" from Mike, since he knew that the presence of light and warmth might affect the natural growth and death rates of the roots he was studying. In the faint glimmer of a 25-watt flashlight, Mike hung a sheet of plastic over one of the large windows that were set into each of the walls. On it, he had traced the path of each of the hundreds of roots that grew pressed up against the glass, marking each segment of line with a date. Mike now proceeded to carefully trace the growth that had occurred since his last measurement. Occasionally he let out a small grunt, which I learned meant that one of his roots had died. Shivering slightly in the alpine cold—since heating as well as light affects root growth—I picked up a marker and got to work as well, occasionally grunting myself. At the end of a year of this tedious but exciting work, Mike had a unique record of the growth and death of the nutrient- and water-apparatus of the forest, something that only a few other forest studies had accomplished. Among the many new insights he gleaned was that the collective length of the root system of a single seedling just six inches in height can be over one mile long.

Nowadays, "minirhizotrons" are more common, for which claustrophobes like me are grateful. Plastic tubes hammered into the ground provide a medium through which a soil scientist can lower a tiny video camera. As it passes through the different soil depths, the camera may capture the dance of a nematode or the squiggle of an earthworm as it records root length. Repeated measurements of the root's length—or disappearance—from the same tubes at subsequent monthly intervals allows the researcher to calculate the rate of root growth—or mortality. In 2001, scientists at the Environmental Protection Agency in Corvallis, Oregon, carried out a review of the research that has been enabled by this approach. The literature indicated that minirhizotrons provide an unprecedented way to study soils, as they afford a nondestructive, in situ method for directly viewing and studying fine roots. Questions that can now be more fully addressed include the role of roots in carbon and nutrient cycling, rates of root decomposition, responses to fertilization, and the significance of interactions between plant roots and soil organisms. Scientists also create sonograms of soil to differentiate solid objects such as roots and rocks in the soil matrix, in ways similar to those that obstetricians use to monitor the growth of an unborn baby in the matrix of amniotic fluid. They also inject radioactive dyes to understand the size and distribution of airspaces within the soil.

The roots and the materials they transport function as complex systems of communication and interchange. For example, the roots of many tree species can graft to roots of neighboring trees, allowing for the exchange of nutrients and water. This phenomenon can also make roots the pathway for the spread of infectious disease, as in the case of the infamous Dutch elm disease (so called because it was first described in Holland in 1921, though the pathogen originated in Asia, possibly China). Caused by the fungus *Ophiostoma ulmi* and spread by a bark beetle, the disease appeared in the United States in 1930. From the point of inoculation, the fungus moves through the xylem—the vascular system of the tree—and quickly reaches the roots. Where elms are planted close together, root grafting may occur, and the fungus can move seamlessly from one tree to the next. Because the American elm is an important tree commercially and aesthetically, plant breeders have been trying to develop a disease-resistant variety.

Grafting to one's neighbor's roots can, however, provide support rather than sickness, as with the Puerto Rican tabonuco tree. Over 60 percent of all stems and basal areas of tabonuco occur in unions, leading to clumps of trees interconnected by root grafts. Researchers have documented that in the tropical lowland forests of the Caribbean, grafted trees grow taller and sustain far less hurricane damage than do nongrafted individuals. These results suggest that a noncompetitive force such as root grafting may be important in maintaining the forest community.

How might this ecological pattern extend to humans? Consider this observation by the writer Howard McCord: "A tree communicates with other trees most intimately and enduringly beneath the ground, as rootlets touch, entwine, and stay. The flickering contacts leaves and branches have with one another in the wind are so random and fleeting. . . . What's important to us goes on *up here*, what's important to a tree goes on generally down below, in a dense, hidden medium we inhabit only after death." In other words, trees symbolically manifest the importance of that which is hidden. Their roots are underground, out of sight, yet they provide support for the tree and are the gathering apparatus for water and nutrients. Tree roots can also symbolize that which we keep hidden from ourselves and others: our troubles, our failings, our addictions, our ill health, and our fears. We know that revealing these hidden roots to our spouses, our friends, our pastor—and most of all to ourselves—is the first step toward finding the strength to overcome them.

TREE FORM AND ITS EXPRESSION OF THE PAST

> The tree is more than first a seed, then a stem, then a living trunk, and then dead timber. The tree is a slow, enduring force straining to win the sky.
>
> —*Antoine de Saint-Exupéry,* The Wisdom of the Sands

The form of a tree is a frozen expression of its past environment and traumas. Just as a fifty-year-old man might limp from a high school sports injury, so do trees carry the physical signs of their history and experience. I know this from my own efforts as a would-be orchardist, which began when my family first moved into our house in Olympia. The fruit trees that we planted have been, to put it generously, disappointing. They were nibbled back badly by deer their first year, and even though we put fencing around them after that, they never recovered fully; their limbs remain asymmetrical and clublike. What explains this long-term memory of trees?

In the 1980s, Claus Mattheck, a German physicist, applied his research on the mechanics of tree structure to the field of arboriculture—the study of tree health and cultivation. Mattheck considers a tree to be a "self-optimizing mechanical structure," a construct that embraces two principles: first, trees make economic use of their resources; and second, they are only as strong as is necessary. If a structure such as a branch is evenly loaded and if all points on its surface withstand the same stress, it will have no overloaded areas (breaking points) and no underloaded areas (wasted material). Elephant tusks, bird wings, human leg bones, and tree limbs are also optimal structures, having evolved to withstand the loads placed on them by the environment. To gain a competitive edge on other plants, trees enhance their ability to survive by sending out branches in all directions, which allows them to present more energy-harvesting leaves to sunlight. To do so, they must develop structures that are strong enough to withstand the forces of gravity and wind. For instance, the taper of a trunk is a tree's response to wind. Wind places the greatest strain on a tree at its base (the "greatest bending moment"), so on windy sites a tree puts on more wood at its base than at the top of the stem. A tree protected from winds in a dense forest usually has very little taper. Branch attachments show similar adaptive behavior. Limbs have collars of extra wood at the point of attachment, so on windy sites, the collars are larger on the windward side of the tree.

When a tree's bark is removed or disturbed, for instance by a person idly carving into it with a knife, the cambium layer just under the bark senses the extra stress on the tree's surface and immediately begins to reduce the stress by growing callus tissue (new wood over the wound). Trees also produce so-called reaction wood in response to the force of gravity.

When a pine tree leans, it grows more wood on the lower side of the trunk to bring it back into an upright position, producing an asymmetrical annual ring. Therefore, sites on trees where extra wood has grown are hints to the arborist of internal injuries or mechanical stress. Sometimes, when I look at our little home orchard or take an evening stroll down a tree-lined street, I think about how tree bodies express themselves. I feel like a combination physicist, psychiatrist, and detective as I piece together the injuries and strains that each tree has encountered in its past and marvel at its ability to overcome such stresses.

THE ART AND SCIENCE OF TREE CLASSIFICATION

> trees are our lungs turned inside out
> & inhale our visible chilled breath.
>
> our lungs are trees turned inside out
> & inhale their clear exhalations.
>
> —*Bill Yake, "inside out"*

TAXONOMIC APPROACHES

Taxonomy is the scientific discipline in charge of classifying and naming living things, based on the similarities and origins of their physical structures. My husband, Jack Longino, is a taxonomist, specializing in the classification of ants. Jack and I met when we were both graduate students, carrying out our dissertation projects at Costa Rican rainforest research stations that were located on opposite sides of the country. I climbed trees in the cool, mist-kissed Monteverde Cloud Forest, while Jack collected ants in the sweltering lowlands of Corcovado National Park. Backdropped by the romantic tropical rainforests we studied, we fell in love quickly and deeply, and alternated visits to each other's field sites for over a year. In one of our most special courtship moments, Jack looked into my eyes and promised to name an ant after me. Nine years later, he presented me with his taxonomic paper describing *Procryptocerus nalini*, a rare ant species that he discovered high in the tropical treetops. He has numerous ants named after him by other taxonomists (it is poor form to name a new species after oneself). His 2005 Christmas present to our children, Gus and Erika, was naming ant species after each of them—*Pyramica augustandrewi* and *Pyramica erikae*—making us, perhaps, the only family whose members each have an eponymous ant. After twenty-five years of studying ants, he has learned how to differentiate hundreds of ant species, tying each one's name to information about its natural history, social interactions, and environment. This ant-colored outlook turns the world around him into a rich tapestry of diversity, whether he is in a remote rainforest or an urban parking lot.

Trees, like ants, can be classified by their taxonomic groups. There are close to ten thousand identified species of trees on earth. Being able to identify them—or even to recognize the ten most common species around us—enriches our understanding of our landscapes. The plant kingdom is divided into a hierarchical set of subdivisions: classes, orders, families, genera, and species. A genus constitutes a group of individuals that share many physical characteristics. Within a species, individuals interbreed and produce fertile offspring. For general scientific use, botanists identify a tree by its genus and species; Douglas-fir is *Pseudostuga menziesii;* bigleaf maple is *Acer macrophyllum.* The official "nicknames" used to abbreviate the names of trees when collecting data or making inventories are made by taking the first two letters of the genus and species: PSME and ACMA.

Dendrologists—people who study trees—have established criteria to differentiate species based on physical appearance. To identify a tree using a botanical key, you break off the end of a tree branch and examine the pattern of its leaves, buds, or flowers. The arrangement of leaves on the stem, for example, may be alternate (individual leaves climbing the stem in alternating steps, with one leaf per node), opposite (two leaves across from each other, both emerging from the same node), or whorled (multiple leaves arranged in a circle, all sharing the same node). The flowers

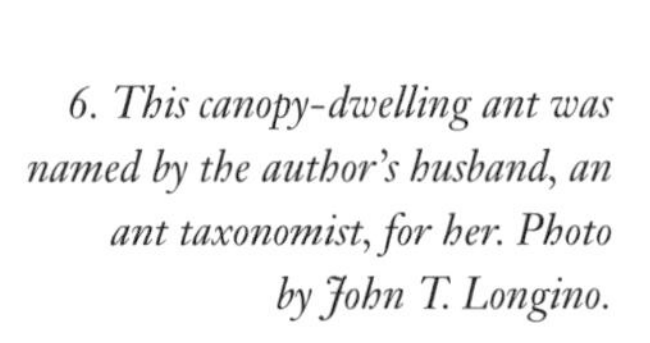
6. This canopy-dwelling ant was named by the author's husband, an ant taxonomist, for her. Photo by John T. Longino.

may have three or five or even twenty-five petals. Fruits may be partitioned into three or five or more sections; the seeds may be winged, enclosed in pods, or entirely naked. The ways in which their structural parts are assembled provide clues for the classification of all the trees in the world. These distinctive characteristics also indicate evolutionary relationships among individuals, with more shared structural elements indicating closer genetic relationships. For example, oaks, in the genus *Quercus*, all house their seeds in the form of acorns. Within that genus, hundreds of different species exist, for example, cork oak, live oak, and valley oak, and they exhibit subtle differences in their physical form—such as acorns that are hairy versus smooth and shiny, or leaves that are lobed versus smooth edged.

In addition to their scientific designations, nearly all trees have common names. These are useful in some ways: they may describe the habitat in which the tree grows (e.g., river birch), how the tree looks (quaking aspen, weeping willow), or some benefit to humans (sugar maple, canoe birch). However, a single kind of tree can have many different names, making it confusing to classify a tree based on its common name alone. *Quercus dumosa*, for example, is called post oak, blackjack oak, and scrubby oak in different parts of the country. In 1908, U.S. Forest Service botanist G. B. Sudworth published a landmark checklist of trees in the United States containing a thousand scientific species names that collectively yielded over nine thousand common names—a vivid demonstration of the multiplicity of common names.

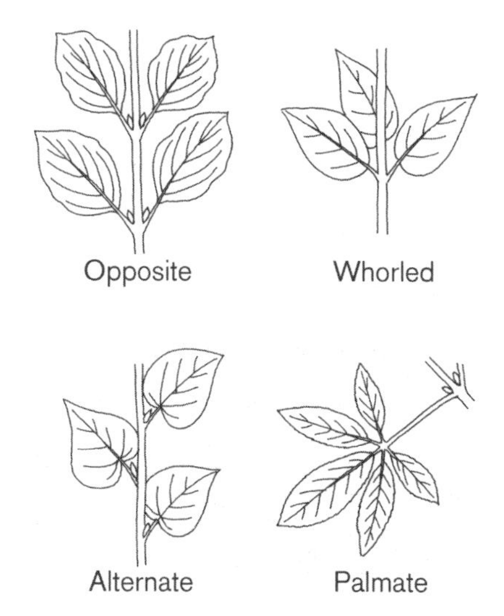

7. Methods of classification of trees and other plants are partly dependent on the way leaves are arranged on their stems. Adapted from John D. Stuart and John O. Sawyer, Trees and Shrubs of California *(Berkeley: University of California Press, 2001).*

APPROACHES TO CLASSIFICATION

Classifying the bewildering diversity of trees taxonomically—whether by common name or scientific name—can be amazingly difficult, even for professional botanists. For many types of forests, especially in tropical regions, no botanical keys are available at all. A one-acre tropical rainforest plot in Amazonian Ecuador, for example, may have as many as 180 different tree species—including some that are as yet undescribed. It can take years, even decades, to learn all the botanical names of the trees that live in an area. Robin Foster, who has worked as a professor, a freelance tropical botanist, an ecologist for a conservation organization, and a research scientist for the Smithsonian Institution, is a heroic figure in the eyes of many botanists. He was, for a while, about the only person who could reliably identify virtually any tree in the diverse rainforests of lowland Central and South America. A soft-spoken man whose genial demeanor belies his sharp eye and strong drive to know every tree species in every forest, he spent years of his life (and a good part of his health, as he has suffered from numerous tropical diseases) learning the subtle differences between species of rainforest trees, relying on his prodigious memory to discern the difference between, for example, *Hieronyma oblonga* and *H. macrocarpa*, two tall tropical trees of Ecuadorean forests, based on the number of floral parts.

Robin, now an adjunct curator at Chicago's Field Museum, is currently working with Steve Hubbell at the University of Georgia to understand the distribution and abundance of trees by mapping and measuring every tree in a 50-hectare (125-acre) plot in a lowland tropical forest on Panama's Barro Colorado Island. Robin had the difficult job of identifying over 235,000 individual trees, from tiny seedlings to giant canopy emergents. The lack of resources for identifying plants in the tropics has been a bottleneck for researchers of tropical ecology and a barrier to public interest for a long time. It has inspired Robin and his colleagues to develop a variety of new identification guides and training materials, which take advantage of digital technology and the vast collections of the Field Museum's herbarium.

Not every location has the benefit of a Robin Foster, however. Because of the shortcomings of a strictly taxonomic approach in places where such expertise is not available, forest ecologists are now keen to establish a different approach—categorizing trees by their overall structure, rather than by a botanical key that depends on ephemeral parts of plants such as flowers and fruits. The goal is to be able to simply look at a tree, recognize how its branches fit together or how its leaves are arranged, and then categorize it according to this gross structure, regardless of its taxonomic group. Humans are adept at classifying objects into structural classes. Go to any sewing store and observe the racks of buttons and zippers, neatly arrayed by size, number of holes, length of teeth, and surface texture. A customer bent on finding just the right fasteners for her soon-to-be-completed dress can locate them efficiently because of the way the shopkeeper arranged them—by structural category.

However, classifying trees with a structural approach is not completely straightforward. The diverse questions and interests that each individual researcher poses require using different pieces and perspectives of the forest. For example, an entomologist (like Jack) who is interested in the ants that march along branch surfaces wants to view the forest from the insect's perspective. An ant would perceive a tree as a set of interconnected transportation networks that consists of a series of two-dimensional planes that she can walk upon no matter what her orientation—upside down, rightside up. Her world is a set of flat surfaces. In contrast, an ornithologist who is interested in the distribution of eagle nests benefits from seeing the forest from a bird's-eye view and would consider the forest not as a series of flat planes, but as an array of static and discrete locations distributed in three-dimensional space—specifically as a set of intersections of branches six inches in diameter or more, and capable of sheltering a good-sized nest. And to a botanist interested in pollination ecology, trees might be viewed as a set of floral "hot spots" for hummingbirds to visit. In their nectar-seeking flights from one flower to another, pollinators connect these dots, either directly, as trap liners, or seemingly randomly, flitting from one point to another. In either case, the distance between flowers is a Euclidean measure related to that three-dimensional volume between flowers. In contrast to the ant, who must traverse the tree branch as a unit, the hummingbird does not care which flower belongs to which branch. For tiny pollen grains, meanwhile, a tree is a giant solid obstacle on which to smack, as they float, nearly weightless, through the shared volume of three-dimensional airspace.

This array of structural possibilities brings to mind a poem from my childhood, "The Blind Men and the Elephant" by J. G. Saxe, which describes six blind men examining an elephant. Each man chooses a different part of the animal to explore—the tusk, the tail, the leg, or the trunk—and each comes to a wildly different conclusion about what an elephant looks like based on his own small sample. In the end, as the poem goes, "each was partly right, and all were in the wrong!" Like the blind men and their elephant, forest ecologists look mainly at the parts. Our ant biologist, ornithologist, and botanist are partly "in the right" to insist on their network, nodal, or volumetric view of a forest, because each does indeed capture an important aspect of a forest's structural complexity.

But they would also be "in the wrong" if they insisted that their view was sufficient to describe forest structure as a whole.

Although as yet no single protocol exists to classify forest structure, several complementary systems are in use. One involves what I call the "Lost Luggage" approach, an idea that came to me on a cross-country airplane trip when my luggage failed to appear on the airline carousel. At the baggage office, the staff person handed me a chart that depicted the various types of luggage, keyed by various structural elements (one handle or two? zipper or no zipper? hard or soft?). Very quickly, I identified the number-coded image that most closely fit my errant suitcase, and almost as quickly, my bag was located in the huge roomful of lost luggage, all of which had been stored by code numbers. Thus, by means of a finite number of structural elements (handles, zippers, hardness), an individual object was identified with ease, just as with the array of sewing goods in the fabric store.

Using this same principle, orchardists have developed a way to classify the different growth forms of their fruit trees in order to impose horticultural regimes. For example, the founders of the Dumfries and Galloway Orchard Network, an organization that encourages sustainable fruit growing in Wales, examined a large number of trees and partitioned them into a small number of categories based on the form of the trunk and branches. They created a booklet depicting the shapes that occur in orchards under varying sunlight exposures and soil types. For each category, such as "fan-trained," "maiden," "bush tree," and "pyramid," they recommend specific pruning and fertilizing treatments. The system works well because they deal with only a small number of tree species (apples and pears) and a limited number of growth forms.

In real forests, however, a nearly infinite roster of tree forms exists. Thus, the lines that separate one tree form from another in orchards are not useful for natural forests. An approach to classifying forest structure was needed. In 1978, a trio of scientists—the French botanist Francis Hallé, British plant anatomist P. B. Tomlinson, and Dutch ecologist

R. A. A. Oldeman—devised a system called "tree architecture." Although they used it to classify tropical trees, the principles of the system apply to all trees. They examined the arrangements of buds and documented the way these units "iterate," or repeat, through time. Anatomically, trees are constructed in a finite number of ways. Some trees put out many buds from a single point at the tip of their branches, while others grow a series of lateral buds. If each bud grows into a branch, that tree exhibits its fullest structural potential—its "true architecture." Hallé and his colleagues determined that all possible growth forms fall into one of twenty-three categories, or architectural models. Leafing through their book on tropical tree architecture is like flipping through an automobile showcase catalogue, with each individual tree species presented in its ideal Platonic form. However, individual trees in the wild—even those of the same species—almost never end up looking like their potential self. They encounter wind, climbing primates, and shade from nearby trees, all of which cause certain branches to fall and some buds to shrivel. So finding model trees with perfect architecture in a rainforest is somewhat like locating a perfect automobile in a used-car lot—possible in theory, but probably not in fact.

A third approach to classifying tree forms occurred to me one morning at a jewelry store where I was getting a necklace repaired. As I waited, I picked up a brochure that described how jewelers classify diamonds. Originally, when there were very few diamond-producing mines, rough stones could be traced to their source by their "signature look." When diamonds were discovered in Canada, Australia, the Congo, and other countries, however, distributors needed a more inclusive and objective system. So they created the "4C" system, in which stones are rated from 1 to 10 on each of the so-called 4C's, four characteristics that all stones possess: cut, color, clarity, and carat (weight). With the ability to describe individual stones by a set of four digits, jewelers seeking a particular form and look of diamond can easily query suppliers.

This system contrasts with the Lost Luggage technique, which sub-

8. These drawings illustrate Hallé, Oldeman, and Tomlinson's system of classifying trees according to their "architecture." Adapted from F. Hallé, R. A. A. Oldeman, and P. B. Tomlinson, Tropical Trees and Forests: An Architectural Analysis *(Berlin: Springer-Verlag, 1978).*

jectively sorts things (baggage, trees) into recognizable groups based on whether or not they have specific features. It also contrasts with the tree architecture approach, which is based on one narrow aspect of arboreal biology—bud arrangement—but does not address subsequent growth dynamics in real environments. Each has its value, however. One of my computer science colleagues, Judy Cushing, and I have been working to develop a new and integrative conceptual framework to categorize trees, using two of these approaches. First we look at general form, following the Lost Luggage model. As in the 4C system, each of the general "representations" we define is then modified by a set of more specific descriptors—such as number of spatial dimensions and whether it is spatially referenced. We hope this will provide a systematic way of describing forest structure that ecologists can use to document not only how forests look, but also how they function and change through time.

DESCRIBING WHOLE FORESTS

Let us shift our perspective from single trees to the whole forest. Except for trees in very dry regions, very cold landscapes, or urban venues, trees occur growing together. A forest stand is much more than a collection of individual trees, just as a city is much more than a group of humans living in the same area. When I visit New York City, I always marvel that something so large and complex can provide a venue for eight million people—and their pets and houseplants, not to mention the wild birds, rodents, and insects that still live there—to awaken, eat, communicate, exchange goods and services, play, and go to sleep. To enter a forest is to marvel at the varieties of trees and their associated plant and animal species, all of them circulating and storing energy, water, and genetic information. The mechanisms used to gain sustenance from soil, rainfall, and air in forests are at least as complex as, and generally far subtler than, the subway systems and office buildings of a bustling urban center. Just as architects and sociologists examine urban physical and social structures, forest ecologists have developed similar approaches to describe how forests are organized.

The definition of a forest in *Webster's Third* is simple: "a dense growth of trees and undergrowth covering a large tract of land." The word's linguistic root reaches into medieval Latin, in which *forestis* means "outside." The term applies to ecosystems as diverse as the sparsely treed seasonally dry woodlands of Latin America, the exuberant rainforests of the Asian tropics, and the monochromatic spruce forests of Siberia's taiga.

Forests have been a major feature of planet Earth for more than 300 million years. The extent of forests has grown and shrunk with climate changes, ice ages, and human population growth and technological innovations. Today, about 30 percent of the world's land surface is covered by forest.

The vast land area that trees cover, the rapid conversion of original forest types to others (usually commercially preferred types), and the difficulty of obtaining information in some countries have until very recently precluded a reliable census of the world's tree population. But lately, that situation has been changing—thanks in part, and somewhat incongruously, to NASA, an agency created to study the presumably lifeless landscapes of outer space. Planetary scientists have invited life scientists into their arena of satellites and supercomputers to help interpret images snapped by satellite sensors of complex patchwork quilts of forests, fields, and urban spaces, connected by rivers. Calibration and interpretation of these images must be coupled with ground-based studies to help foresters inform land use managers, conservationists, and policymakers about the state of the world's forests. As a result, some of the finest forest ecology studies being carried out today are the result of NASA-funded multidisciplinary collaborations.

With these emerging tools, geographers have been able to make reasonable inventories of the world's trees. Using such diverse sources as tree counts by rural forestry departments and remote imagery supplied by NASA, foresters at the Food and Agriculture Organization, based in Rome, produce estimates of tree cover every five years. In 2005, they reported the area of forested land as 3,952 million hectares, or 15,258,853 square miles, an area equal to the United States and Australia combined. Coupling those numbers with estimates of tree density, they calculated the total number of trees on Earth to be about 400,246,300,201. I wondered how many trees that would be per person—or perhaps it would work out to how many people per tree. I looked up the world's human population and learned that as of December 31, 2005, humans are

6,456,789,877 strong. Punching the figures into my calculator, I calculated that the world supports sixty-one trees for each person on Earth. When I think of the millions of people living in the densely populated and virtually treeless urban landscapes of Mexico City, Tokyo, and New York City, that figure seems very large to me—sixty-one trees for each person walking through Times Square! But when I told my husband, Jack, about this tree-to-human ratio at breakfast, he reflected for a moment and then voiced wonder that the ratio was so small. "Each person gets only sixty-one trees in a lifetime?" he mused over his muffin. "That seems hardly enough to supply just the firewood we'll use in our woodstove for the next few winter seasons, let alone the lumber that's in our house and the paper I put through my printer." His reflections made me revise my original impression, and reinforced the sense that I need to think about ways to look after my sixty-one trees, wherever they might be growing in the world.

CLASSIFICATION OF WHOLE FOREST STRUCTURE

Foresters do more than count trees. They have also developed ways to classify whole forests by their structure—how they are put together, their shapes and collective forms. Just as a zookeeper might herd together all of the four-legged mobile animals and put them in a pen, and then guide all of the slithering legless animals into a secure glass cage, so do foresters rely on the salient structural elements of forests for classification. Geoffrey (Jess) Parker, an ecologist at the Smithsonian Environmental Research Center in Edgewater, Maryland, has for decades been working to understand the structure of the aboveground parts of forests, both in his native eastern deciduous forests of Maryland and in other forests around the world. His formal definition of forest structure is "the organization in space and time, including the position, extent, quantity, type and connectivity of the aboveground components of vegetation, as well as standing dead trees ('snags'), fallen logs, and the open spaces between canopy elements." He notes that many terms have been used interchangeably in reference to forest structure, but they emphasize different aspects. Those who study *forest physiognomy*, for example, focus on the shapes of individual crowns. *Forest architecture* describes the growth patterns and resultant forms of stems. *Forest organization* implies the statistical distribution of forest components in space or time, and *forest texture* refers to the sizes of crown units composing the overstory, apparent only from above the stand.

Ecologists such as Jess recognize the central place that structure plays in many critical forest attributes. The arrangement and vertical distribution of leaf area within tree crowns, the distribution of branches around the trunk, and the distribution of trees of different heights can have a strong effect on microclimate in the forest understory, forest functions such as photosynthesis and respiration, the cycling of water and nutrients, rates of growth, the diversity of understory plants, and suitability for wildlife. An abundance of snags, for example, greatly increases the number of nest sites for cavity-nesting insects, mammals, and birds such as woodpeckers and resplendent quetzals. Primary and secondary cavity-nesting birds make up a substantial proportion of the avifauna in both tropical and temperate forest ecosystems.

FOREST AIRSPACE

Until recently, structural classification systems focused on the arrangement of solid objects within a forest—the trees, stems, and foliage. But recent work has involved measuring the "negative space" as well, that is, the volume not occupied by solids. Dr. Roman Dial, a fit, bearded researcher from Alaska Pacific University who completes triathlons as easily as he writes database routines, quantitatively describes the airspace of world forests in order to better understand their structure. Roman first shoots off a single long rope to connect two trees, which serves as his horizontal aerial transect. The height of the line depends on the forest in which he works, but Roman almost always works in tall, old-growth forests where his horizontal transect stretches over two hundred feet above the ground. He then hangs a set of vertical ropes at five-yard intervals along the transect, to create his sampling grid. Attaching himself in his climbing harness to the top of the first dangling rope, he uses a laser rangefinder to measure the distance from his eye to the nearest solid object (a branch, a leaf) in eight compass directions around himself. He then lowers himself twenty feet down the vertical rope, and again shoots in all directions. He repeats this at twenty-foot intervals down that rope and from all the other ropes hanging along the horizontal transect line. From these data, he calculates the volume and shape of the airspace of that chunk of forest. His work, which leads ultimately to graphic representations that are both stunning in their beauty and intriguing in their meaning, allows us to visualize a bird, pollen grain, or pollutant particle interacting with the interstices of the forest. With Roman's images, we can better see the possibilities for airborne entities in the forest: they can move around, crashing into or settling gently on the surfaces of the solid canopy that they encounter. This is a first step in getting a new and integrative understanding of how all the elements of the forest interact.

Roman's work can be extended into other areas. I once described his airspace research at a public lecture for gardeners, which was attended by a Buddhist monk-in-training. After the lecture, he told me how Zen arts epitomize the relationship between form and emptiness. In painting and calligraphy, he explained, empty space is as important as pigment and lines. Start with a blank piece of paper. If an artist paints on that sheet of paper, say, a small bird on bamboo gazing out over an infinite horizon, everything changes. Now you have form: the bird, the bamboo, the horizon. And you have emptiness, as the bird's gaze draws your eye to the vast expanse beyond the horizon. Only out of form does emptiness become possible—and vice versa. Every object has both form and emptiness—a painting, a forest, a human being. Even major evolutionary processes can be affected by forest structure. For example, the gliding habit of a variety of unrelated arboreal animals—flying squirrels, gliding lizards, flying snakes—has been influenced by the distances between large trees.

A MATTER OF PERSPECTIVE

It is foolish
to let a young redwood
grow next to a house.

Even in this
one lifetime,
you will have to choose.

That great calm being,
this clutter of soup pots and books—

Already the first branch-tips brush at the window.
Softly, calmly, immensity taps at your life.

—Jane Hirshfield, "Tree"

We measure trees and forests both absolutely, in terms of girth, height, and volume, and relatively, by scaling them against human size. Large trees are mysteriously riveting. More than a few of my canopy research colleagues sporadically leave excited messages on my voicemail: "Hey! I touched the top of a 293-foot-tall tree today at noon!" But even ground-bound people find extremely big trees fascinating. For example, we grant a special status to the tallest living tree in the United States. In 2002, the title of tallest tree was awarded to a coast redwood named the Stratosphere Giant, measuring 369 feet, five stories taller than the Statue of Liberty. Four years later, however, that title was challenged by three other coast redwoods, and the championship for tallest tree was finally settled on a tree called Hyperion. Named for a Titan, the son of Gaea and Uranus and the father of Helios, Hyperion is 379.1 feet in height. It was discovered by amateur naturalists in Redwood National Park during the summer of 2006 and is now accepted to be the world's tallest living thing. I revere these extremely tall trees, in their lofty grandeur, as examplars of Herman Hesse's "penetrating preachers"—trees that "stand alone . . . like lonely persons. Not like hermits who have stolen away out of some weakness, but like great, solitary men, like Beethoven and Nietzsche."

The tree with the largest volume is the Del Norte Titan, discovered in June 1998 in Jedediah Smith Redwoods State Park, California, by Humboldt State University ecologist Steve Sillett and naturalist Michael Taylor. This tree has an estimated stem volume of 1,366.4 cubic yards and is 306.98 feet tall, with a diameter at breast height (4.5 feet) of 23.68 feet. Its mass is equivalent to fifteen adult blue whales, the largest animal on earth. Each year, this tree produces enough new wood to make a ninety-foot-tall tree with a trunk twelve inches in diameter. If all of the Del Norte Titan were cut into boards one foot wide, twelve feet long, and one inch thick, the line of planks laid end to end would stretch over a hundred miles and could build 120 average-sized houses.

Many other trees give us plenty to ooh and ahh over. For example, consider the great banyan in the Indian Botanical Garden of Calcutta, whose canopy covers an area of three acres. Or the aspen, which forms large clonal colonies of genetically identical trees in the northern Midwest of the United States. Because new stems arise from root sprouts originating from the parent tree, what appears to be a large grove is actually a single individual. One such grove covers thirty-five acres and has a mass of 6,600 tons.

Although the fascination with extremely large trees is understandable, I consider them more as freaks on the midway of the arboreal world than the living things I relate to daily. What intrigues me more is trees' effects on and interaction with our physical world. Several years ago, I worked on a scientific project measuring throughfall (the amount of rainfall passing through the canopy that reaches the forest floor) and stemflow (the amount that slides down the branches and trunk). I spent weeks analyzing the data and then teasing apart the structural features of the canopy (foliage? twiglets? bark texture?) that determine how, and how much, rainfall finds the ground. At one point, I leaned away from my computer screen and shrunk myself in my imagination to the size of a single raindrop. I then envisioned myself participating in the dance of the raindrops as they fell through the canopy, ricocheting, fragmenting, coalescing, turning into tiny bubbles, and finally disappearing into vapor.

Human beings tend to see trees as single entities, scaled to their own needs. But what about the fascinating array of other organisms that interact with them? I did some calculations to arrive at an answer of sorts—at least for an ant's-eye perspective. First, imagine a *Procryptocerus nalini* ant, one-twentieth of an inch long, walking along a branch to gather pollen and spores to bring back to her arboreal nest. This corresponds to me, a human Nalini about five feet five inches tall, walking along a road to shop for, say, pasta and tomato sauce for tonight's dinner. Except that if I were to scale my road up to be equivalent to the branch the little ant is on, ten inches in diameter and fifty feet long, it would have to be a thousand feet wide and twelve miles long. We experience another major difference in how we navigate our surroundings as well. Because the ant weighs so little, she is not restricted topologically to the upper surface of the branch, as we humans are, but can walk along its underside; indeed, she can encircle its entire cylindrical surface. For me, this would equate to tying on Velcro shoes and casually stepping out to stroll along the underside of a suspension bridge that arches over the vast empty space below it.

Another question of scale concerns the distance between branches—a concern especially for epiphytes, such as orchids, ferns, and bromeliads, that have evolved in the canopy, deriving structural support but not nutrients from their host trees. When epiphytes fall to the ground, the lack of light, the greater relative humidity, and the presence of terrestrial pathogens cause them to die within months. Although their niches in the canopy provide access to the right amounts of sunlight and rainfall, when it comes to reproduction they face a perennial challenge. Whereas trees merely need to create seeds that will fall or be blown to a site on the forest floor where they can successfully germinate, epiphytes need to get their seeds to a suitable spot on another branch high in the canopy. To do so they must rely on the wind or on flying mammals and birds—and a great deal of luck, given the widely dispersed branch space amid vast amounts of airspace in that high treetop realm.

Let's rescale the forest again, this time using measurements made in the cloud forests of Costa Rica. A typical adult epiphyte is four inches in height, the width of a typical branch is six inches, and the distance to the next branch is about sixty feet. Extrapolating to human size, the equivalent distance to the next branch over is nearly one thousand feet. Thus, if you were to accomplish what a canopy bromeliad must do to get its seeds to the next available safe spot, you would have to fling a BB across three football fields and have it land on a surface eight feet wide. Considering what is required, it is remarkable that any bromeliad successfully reproduces at all. It also explains why many canopy-dwelling plants make huge numbers of spores and seeds. Certain species of orchids, for example, produce seedpods that hold exceedingly small, dustlike seeds, as many as 400,000 per capsule. The sheer number of seeds escaping and being blown through the canopy ensures that at least a few will beat the odds and land on a hospitable site to start the next generation.

DYNAMICS OF FOREST STRUCTURE

Contrary to the image we often hold of their being ancient, timeless places where time stands still, forests are dynamic. On a blustery day in the windy season, I walked out to my study plot in the tropical cloud forest of Monteverde, Costa Rica. Three miles up the trail, one of the largest trees in the plot had toppled, an individual I knew intimately from scores of trips up and down its trunk. I had named this tree "The Mansion" because its giant crown suggests a gracious home of many rooms, each branch festooned with orchids, bromeliads, and ferns and inhabited by a diverse assemblage of arboreal animals. But the tree had fallen in the previous night's windstorm, tearing open a huge gap in the fabric of the forest canopy, which created a sunlit area on the ground that had previously been held in deep shade. I took a seat on the toppled trunk, feeling sadness at its transformation from vertical to horizontal, living to dying. Yet the sunlight that could now penetrate to the previously dark forest floor would, I knew, awaken the dormant seeds that lay buried in the soil at my feet. In a few months, new tree seedlings would sprout and make their own way to the canopy. By the time I stood to leave the light gap, I had shifted my saddened attitude, viewing the trunk of the fallen tree now as a fallen hero who inspires his followers to move from stasis to action, from darkness to light.

Techniques for measuring how forests change through time are well established. The traditional tools of a timber cruiser—a forester who measures the location and volume of trees in a forest—are simple and inexpensive and can fit into a small backpack. The cruiser measures a tree by wrapping her measuring tape around the trunk at a level exactly 4.5 feet above the ground, to arrive at "diameter at breast height," or DBH. Special "D-tapes," calibrated in units of pi, allow the cruiser to read the diameter directly. Height is worked out by triangulation: standing a measured distance from the base of the tree, the forester looks through a clinometer, a small metal box that measures the angle from her eye to the tip of the tree. Using trigonometry, she then calculates the distance to the top. At the end of a single day, a cruiser will have generated enough information to report on the average height and volume of all the trees, and their variation, thereby creating an accurate static picture—a snapshot—of the forest stand.

Documenting the dynamics of the stand—making a *moving* picture of the forest—involves more work. To do this, foresters establish permanent plots, round or rectangular, to which they return at intervals of months and years. They then measure the distances and angles of trees relative to fixed reference points. Very few humans can remember what individual trees look like for more than a few months, and so researchers identify each tree by nailing a quarter-sized aluminum tag to the trunk. Attaching a tag to a hitherto anonymous tree is like giving a new puppy a name—it becomes known to us as an individual, and information on it can be retrieved at any time. Tagged trees are remeasured at intervals of a year, five years, or a decade. With these data, an ecologist can estimate the longevity of the trees and the replacement time for a whole stand.

Without the tags, and the ability to follow individual trees and parts of trees that last far longer than the lifetime of a single field season or the spatial memory of a researcher, we would not be able to understand the basic dynamics of the forest. Tags are such a ubiquitous tool in the measuring of trees that ecologists use them without thinking. Although no studies have revealed that nailing a tag into a trunk damages the tree, tags can, however, have a profound effect on the forest, one that I had not considered until I went tree climbing with a conceptual artist named Bruce Chao. Bruce is chair of the glass-blowing department at the Rhode Island School of Design and a creator of giant and thought-provoking three-dimensional installations that he places in forests of New England. A few years ago, Bruce became interested in the forest canopy as a medium for art. In 2003, he joined my colleagues and me when we were mapping trees for our studies of forest structure. We worked in an old-growth site, where five-hundred-year-old Douglas-fir trees pierced the sky to heights of 250 feet. Branches were muffled in soft green moss that was decades older than my parents. We noted the location of every branch of all the trees in our study sites to get baseline information that we and other ecologists could use. This involved rigging tall conifers with ropes, ascending the ropes with climbing harnesses and Jumars (mechanical ascenders), and hauling up a laser rangefinder to measure the x, y, and z coordinates of each intersection of branch and trunk. Every five meters up the trunk, we nailed a round aluminum tag to indicate our height above ground. Placement of those tags would help us readily find branches again in the three-dimensional world of the tree crown when we returned to remeasure our plots.

Bruce stayed in the tree all day with us, listening to us shout numbers to each other, speculate on which tree had the greatest volume of wood, and argue over which trees we would include in our personal Top Ten Tree Species lists. At day's end, we rappelled to the ground and repaired to the campfire. I asked Bruce if he had been inspired to create a piece of art by anything he witnessed that day in the canopy. He paused. "To me," he said in his quiet voice, "the most interesting sight was the tags." The tags? He explained that the presence of the tags permanently changed the nature of the stand of trees. Each of the trees we had climbed had a faint but distinct "trail" of small silver dots going up the trunk, signifying that humans had been there and that they would come back. The tags, he said, gave a ping of the past and the future to the present. He was neither offended nor repulsed, but simply affected by the handiwork of humans in something that was previously only forest. His comments made me realize that all of our access techniques, which we so glibly call "nondestructive," have an effect on the forest, even if it is just the sound of a Jumar clinking against a carabiner or the glint of a tag against bark.

TREES AND FOREST FORMS IN OTHER REALMS

Trees help you see slices of sky between branches
Point to things you could never reach.
Trees help you watch the growing happen,
Watch blossoms burst then dry,
See shade twist to the pace of a sun,
Birds tear at unwilling seeds . . .
A tree is a lens,
A viewfinder, a window.
I wait below
For a message
Of what is yet to come.

—*Rochelle Mass, "Waiting for a Message"*

TREES AS RIVERS, RIVERS AS TREES

From three hundred feet above the forest, individual trees blur together, becoming a landscape seen by the eye of a raptor or the eye of a hurricane. From this vantage point, too, rivers become trees that have fallen, flattened, and filled with water. I have always been a bit jealous of hydrologists (people who study streams and rivers) because they have developed consistent ways of mapping stream courses and predicting their movements, a capacity that exceeds the ability of forest ecologists to create maps of individual trees. Drawing on the field of morphometry (the measurement of the shapes of things), two researchers, Robert Horton and Arthur Strahler, established a classification system in the 1940s and 1950s based on the hierarchy of stream segments. The main stream of a

9. Streams and rivers are "dendritic" in form, structurally analogous in many ways to trees and other objects in nature. Photo by Greg and Mary Beth Dimijian.

river, or "main trunk," receives materials from its tributaries, just as tree trunks receive moisture from rainwater flowing over their upper branches. Stream channel segments are ordered numerically from a stream's headwaters to a point somewhere downstream. Numerical ordering begins with the tributaries at the stream's headwaters, which are assigned the value of 1. A stream segment that results from the joining of two first-order segments is given a value of 2. Two second-order streams form a third-order stream, and so on. Horton and Strahler were interested in spatial properties that relate to the entire stream system, and deduced that the ratio between the number of stream segments in one order and the next—called the bifurcation ratio—was consistently around 3, which is the same ratio as between the root system and branch structures of trees.

Other natural branching networks have patterns similar to this stream order model, including, somewhat surprisingly, certain marine animals. Coral reef ecologists, for example, use stream-based analyses to understand how corals grow. *Acropora*, or staghorn coral, dwells on the Great Barrier Reef in Australia. Common in shallow water, it is successful because individuals of this genus have light skeletons that allow them to grow quickly and overcome their neighbors, much as the weeds in my garden rapidly outgrow my sluggish tomato plants. *Acropora* has a distinctive way of branching. As its central axis increases in length, it buds off smaller radial branches. Using Horton and Strahler's methods, researchers determined the bifurcation ratios of the staghorn corals and found them to be similar to those of streams.

TREE FORMS AROUND AND WITHIN US

Tree structure extends to the things we use in everyday life. Plastic is made up of long-chain molecules called polymers, which in turn are composed of many smaller molecules, called monomers, that are covalently bonded together. About 60 percent of the chemicals used by the plastics industry are used to make polymer products, including nylon, film, and kitchen countertops. Specialized polymers result when the molecules are oriented in specific ways. Many small side groups along the polymer chain, for example, create more restricted movement and a very strong structure; Kevlar, which is used for policemen's bulletproof vests and windsurfing sails, is one such polymer. Another specialized polymer is the dendrimer, whose name indicates its treelike structure. Here, branches keep growing out of branches, and more branches grow out of those branches. This shape lends the polymer unusual properties. One silicon-based dendrimer can trap oxygen molecules in its branches, making it a candidate for the production of artificial blood.

A wide variety of objects in nature, too, follow these patterns and can be quantified, visualized, and understood in terms of trees. These include blood vessels, the trachea of our lungs, neural pathways, cave tunnels, grounded lightning strikes, trail systems, and even the microcrystalline formations of frozen water. Recently, my family and I visited the "Bodies" exhibit at a museum in Seattle. This popular show invites visitors to examine the intricacies and complexities that lie beneath our very own skin. The use of a polymer preservation process allows the living to view the dead from remarkable perspectives, with each room focusing on a different body system: skeletal, digestive, respiratory, integumentary. Far and away, the favorite room for each member of my family was the one devoted to the circulatory system, in which the delicate patterns of arteries and veins were depicted in bright red and blue. I gazed for half an hour at the structure of capillaries in the lungs. These were clearly trees that branched out into the tiny alveoli of our breathing apparatus. The heart revealed the broad trunk of the aorta, which bifurcated into the separate trees of arteries and veins within the auricles and ventricles. I could as well have been mapping the branches of trees in my study site in Costa Rica, but instead I was looking into the very core of a human who, perhaps like me, held trees in her heart.

ZEN GARDENS AND INVISIBLE TREES

Some remarkable work relating to tree structure and its power over the human psyche has been done in Zen gardens. For centuries, monks have explored life metaphors in the carefully raked and tended areas around their temples. Ryoanji Temple in Kyoto, for example, has been a Zen place of worship and meditation since its construction in the 1450s. The temple garden is a rectangle surrounded by earthen walls on three sides, fronted on the fourth by a wooden veranda. Inside the rectangle is an expanse of white pebbles and fifteen rocks of various sizes, arranged in five separate groupings. The white pebbles are raked every day, with perfect circles around the rocks and perfectly straight lines in the rest of the space. Over the centuries, various explanations for the garden's layout have been suggested, such as that the white gravel represents the ocean and the rocks the islands of Japan, or that the rocks represent the Chinese symbol for heart or mind. In any case, there is no doubt that the site exudes a strong sense of serenity, as many thousands of worshipers and tourists can attest.

In 2002, visual-imaging scientists Gert van Tonder and Michael Lyons published a series of papers that revealed a possible explanation for the strong power of the garden. Applying a shape-analysis technique that revealed hidden structural features in the garden's empty space, they theorized that it is the empty space created by the placement of the rocks, rather than the rocks themselves, that has so intrigued visitors over the centuries. Studies of how humans and other primates process visual images suggest that we have an unconscious sensitivity to the "medial axis" of shapes—that is, our mind subconsciously inserts lines to connect the edges of the shape. At Ryoanji, van Tonder and Lyons concluded, when a visitor looks at the clusters of rocks on the expanse of white pebbles, he is subliminally able to discern the image of a tree—its trunk and branches. The accompanying feeling is one of calm peace.

TREES, RELATIONSHIPS, AND ABSTRACT THINKING

Trees give us ways to think about relationships. Genealogy, the study of how families have descended from their ancestors, depicts individuals as living on family trees. Dozens of software programs and companies have been developed to help identify lost branches of people's family heritage. More recently, this effort has been expanded, and rather than mapping a single human family, we are now attempting to map the family that is all of life—all 1.7 million known species. The Tree of Life, as the project is known, traces the patterns of descent of life over millions of years to one common ancestor. Groups that are more distant from each other evolutionarily have more branches between them. Some branches are longer or shorter than others, indicating whether they are distantly or closely related. It is a collective international project involving supercomputers and hundreds of scientists, from beetle taxonomists to molecular geneticists to database engineers, whose goal it is to find patterns of relationships that best account for the DNA sequence data. Once the project is completed, scientists will have a framework for incorporating newly discovered species into the tree of life and a better understanding of the patterns of biodiversity. Despite data gaps and tremendous computational challenges, the building blocks have an elegant simplicity, thanks to the underlying structure of trees.

Tree forms permeate the way we think and understand in other abstract modes as well. The evolution of human language, for example—its families and its dialects—is typically described in terms of a tree. Here again, evolutionary proximity is graphically portrayed by the positioning and density of branches, with the Finno-Ugric languages, say, which are spoken in scattered areas from Finland to Siberia, taking up a relatively small part of the tree's crown relative to the Romance languages. Philosophy and computation, too, rely on trees as models. For example, intuition and analogical reasoning use the tree structure in applications as diverse as school grading systems, musical analysis, and establishing pathways of information dissemination. Branching models are used to represent networking theory, web design, and application development for artificial intelligence and robotics.

Tree forms also surface in medicine, as I learned a few years ago when my brother Mohan, a specialist in internal medicine, invited me to speak to a group of medical residents about how trees ameliorate human health. When I joined Mohan on his rounds, his students alerted me to another link between trees and medicine, the use of "decision trees" to maximize the accuracy of their diagnoses. These trees—in the form of a Web page or printed document—are used to analyze a particular set of symptoms, leading a doctor to a decision that is informed by thousands of preceding cases. Let's say a patient comes to Mohan complaining of chest pain. Guided by a published decision tree for that symptom, Mohan will ask questions that lead him to decide whether the pain originates from a cardial cause or an intestinal one. He starts with the first of two bifurcating branches in his tree: Is the pain sharp or is it tight? Does the patient burp, or not burp? At some branches, specific tests, such as an EKG, are ordered, and the results direct the further branching of the diagnostic tree. Finally, the doctor arrives at a "leaf," or the terminal point of a series of branches, which articulates the diagnosis and the optimal treatment. The very structure of the tree, with its enforced pauses at each diverging pathway, allows the doctor to think carefully about the diagnosis at multiple points along the way.

These models function as points for reflection. Robert Frost, in his evocative poem "The Road Not Taken," describes his own decision to follow one branch in the path of life, rather than another.

> Two roads diverged in a wood, and I—
> I took the one less traveled by,
> And that has made all the difference.

I think back to those afternoons of my childhood, the hours spent learning how to cross from one branch to another in my favorite maple tree, BigArms, as well as the moments when I would pause in my arboreal acrobatics to imagine the different pathways I could take on my way to becoming a grownup. Each idea would start in my mind as a single point, a specific vision of what I wanted to be—a doctor, a dancer, a veterinarian, a cabin girl on a sailing ship—and then my imagination would branch into the multiple directions I would travel in the future to make that dream happen, branches as complex and ordered, as definite and mysterious, as the twigs at the tops of my tree.

Chapter Two

GOODS AND SERVICES

> We have nothing to fear and a great deal to learn from trees, that vigorous and pacific tribe which without stint produces strengthening essences for us, soothing balms, and in whose gracious company we spend so many cool, silent, and intimate hours.
>
> —*Marcel Proust,* Pleasures and Regrets

Fanning out through the busy corridors of our town's shopping mall, twenty-five college students in my "Trees and Humans" class set out on an afternoon quest: to find and record any items, besides food, that were made from or with the help of trees. Two hours later, they coalesced at the indoor fig tree, a tiny island of nature in this vast ocean of consumer goods, and compared notes. As each student read his or her tally of tree-derived objects, I wrote them on a large pad of paper. An hour and many pages later, our list encompassed an astonishing array of things that trees provide. Some of these items had obvious arboreal origins, such as baby cribs and baseball bats, barrels and landscaping bark, birdhouses and blocks, books and benches, crutches and coffee filters, guitars and grocery bags, and pencils and pine oil. Others were more surprising: adhesives, animal bedding, antacids, artificial vanilla flavoring, automobile instrument panels, bagpipes, baking cups, bassoons, beds, billboards, birthday cards, broomsticks, bulletin boards, buttons, candy wrappers, carpet backsides, CD inserts, ceiling tiles, charcoal, chewing gum, colognes, computer casings, cooking utensils, cork, cosmetics, crayons, decoys, diapers, disinfectants, drumsticks, drywall, dye, egg cartons, electrical outlets, eyeglass frames, fireworks, fishing floats, food labels, football helmets, fruit pie filling, game boards, garden stakes, golf balls, golf tees, gummed tape, harmonicas, ice cream thickener, imitation bacon, inks, insulation, kites, lacquer, linoleum, luggage, magazines, medicines, menthol, milk cartons, movie tickets, nail polish, particle board, party invitations, photographic film, piano keys, Ping-Pong balls, pitch, plates, postage stamps, posters, price tags, puzzles, rayon, rubber gloves, sausage casings, seesaws, shampoo, shatterproof glass, shoe polish, sponges, stain remover, stepladders, stereo speakers, syrup, tambourines, telephone books, tires, toilet paper, toothpaste, turpentine, umbrella handles, vacuum cleaner bags, varnish, vitamins, waxes, xylophones, and yo-yos.

After we'd all marveled at the length and diversity of our list, one student presented a short report on our annual consumption of trees. The average American, he said, uses over a ton of wood each year, equivalent to over 43 cubic feet of lumber, 681 pounds of paper, or a single tree one hundred feet tall and eighteen inches in diameter. Another way to view our use of wood is in terms of a cord, which is a pile of wood four feet high, four feet wide, and eight feet long. This amount of wood produces 7,500,000 toothpicks, 61,370 business envelopes, 4,384,000 postage stamps, 460,000 personal checks, 1,200 copies of *National Geographic*, 250 copies of the Sunday *New York Times*, or 12 dining room tables large enough to seat one host and seven guests.

Overwhelmed by these statistics, we revisited a discussion we had had earlier of Abraham Maslow's "hierarchy of needs" (see the introduction), in an attempt to organize the many relationships that trees and humans share. To discuss each entry in the vast world inventory of tree-derived products—over five thousand of them—would require many books and years of study. Instead, in what follows I will focus on the most basic items on that list—those that fulfill the physical needs of human beings: food, clothing, and other critical objects that have economic value.

THE BASIC BUILDING BLOCKS

The pathway by which tree-derived products move from landscape to factory to market shelf starts with those who select, cut, and haul trees from the forest to the mill: the loggers. The stereotypic woodsman, with his chainsaw, caulk boots, and can of snuff, has often been vilified as the hatchetman of the forest and the enemy of environmentalists. Stereotypes embody but partial truths, however. In the 1970s, I worked in logging camps in southeast Alaska, flying out to remote camps pitched in the shadows of ancient hemlock and spruce trees. I met many loggers there who possessed a deep knowledge and respect for the trees they felled—although they expressed these in different ways than did the conservationists I encountered at forest preservation rallies. Standing together in the predawn dark, waiting for the sky to lighten before jerking their chainsaws awake, my companions spoke in a language that might be considered terse—but no more so than that of the poet Gary Snyder, a former logger and choker-setter himself:

Why Log Truck Drivers Rise Earlier Than Students of Zen

In the high seat, before-dawn dark,
Polished hubs gleam
And the shiny diesel stack
Warms and flutters
Up the Tyler Road grade
To the logging on Poorman Creek.
Thirty miles of dust.

There is no other life.

There may be no other life for the logger, but there is a world for the tree to enter after the logger's job ends—when the tree he has cut falls to the ground and the cable drags it up the hillslope, where it is loaded on the logging truck that takes it to the mill. A cut log arriving at the mill is like the proverbial pig, all of whose parts—except the squeal—are used by the farmer. Bark, which makes up 20 percent of a tree, is the first material to be removed from a log. Although the dead outer skin of the tree might seem like a waste product, it is a source of resins, waxes, adhesives, mulches, and oil-spill control agents. Its primary importance, however, is as "hog fuel"—a "hog" being a mechanical shredder or grinder. Ground bark is then burned in giant furnaces, which heat water to create steam to fuel the mill; it provides over half of the mill's energy, reducing reliance on fossil fuels and such pollution-generating fuels as used tires.

The end of a cut log reveals two nested circles of wood: the outer rim, or sapwood, and the darker inner core, or heartwood. Sapwood is the younger wood, made up of living cells. It conducts water from the roots to the leaves and transports the energy harvested by the leaves via photosynthesis down to the roots for storage. As a tree increases in age and diameter, an inner portion of the sapwood becomes inactive and then stops functioning altogether, forming the inert and dead heartwood. The term *heartwood* is something of a misnomer, since trees can thrive even if their hearts are completely decayed. The inner heartwood of some old trees, however, can remain sound for hundreds or even thousands of years, impregnated as it is with chemical compounds that make it resistant to fungal decay.

Over the past half century of modern milling, humans have learned to create standard types of lumber to make their homes and furniture, mainly from the heartwood. No matter how one feels about logging and clearcuts, it is impossible to stand in a lumber mill and not feel awe. Vast in size, cavernous in feel, and minimal in aesthetics, a lumber mill is a maze of catwalks, conveyor belts, and five-story-high circular saws that render huge tree trunks into orderly stacks of two-by-fours at the end of the run. The noise level is extreme, and ear protection and hard hats are required, even for visitors. When I watch a mill in operation, one of the most astonishing attributes is the paucity of human workers inside. Sometimes only three or four people are on shift, overseeing the processing of literally thousands of board feet a minute. How is this possible? The saws, conveyor belt speeds, and cutting designs—tailored to the shape and size of individual logs—are all managed instantly with computer "eyes" and "brain." A single employee "in the cab" watches over things, keeping his eye not on the lumber but on video monitors trained on machines and stations all over the plant. This lumber lifeguard is ready in an instant to stop a belt for a board gone awry or to change the saws when they need sharpening. The computer can maximize the length and width of lumber each log can yield, position the log with mechanical shovers, and then—*zzzzzzzzzzzip*—the blade does its rapid work and the products we know from the hardware store settle out: two-by-fours, four-by-sixes. The scrap wood falls into the moving waste bins to be made into chips for pulp (and thence paper), lawn dressing, and fuel to power the mill itself.

Then there is plywood, layers of wood ("veneer") joined together by layers of glue. The conversion of roundwood (the trunk that arrives without branches or roots at the lumber mill) into veneer and finally into plywood is efficient and overseen by computer technology as well. After its bark has been stripped, the naked log is mounted on what looks like a giant paper towel dispenser, and the peeler—a long, incredibly sharp blade—is placed against the side of the rotating log. Layers of veneer, cut to thicknesses ranging from one-fortieth to three-eighths of an inch, unwind from the log in long, flexible sheets. They then go to the dryers before they are glued. Driving off the moisture rapidly, to a moisture content of less than 10 percent, ensures minimal warping and maximal strength. Most dryers are long chambers with rollers on belts that move the veneer through the chamber. The majority of high-temperature (above 212°F) veneer dryers depend on steam as a heat source, which is most often generated by burning waste wood from the mill itself. After drying, machines stack alternating layers of veneer that have been spread with glue by rollers or sprayers. These are then subjected to pressure (75 to 250 psi) for a few minutes to many hours to bond the different veneer layers in intimate contact. Finally, like the lumber created from the tree's former forest mates, the plywood is packaged and trucked to hardware stores and construction sites around the country.

Wood itself is made of tiny fibers, including cellulose (a material in plant cell walls, consisting of long chains of sugars) and hemicellulose (short, branched strings of sugars). Cellulose is crystalline and strong, whereas hemicellulose has a random structure with little strength. A third component, lignin (a large, cross-linked molecule), fills the spaces in the cell wall between cellulose and hemicellulose, and serves as the natural glue that holds them all together—the peanut butter between the bread slices. Lignin, which tends not to degrade, functions as an effective barrier against the bacteria and fungi that can decompose plant tissues. When wood is turned into pulp for paper, heat and chemicals dissolve the lignin and release the cellulose and hemicellulose fibers, which leads to a plethora of goods for humans. For example, nitrocellulose and other pulping by-products are used for rocket propellants, explosives, cleaning compounds, deodorants, artificial vanilla flavoring, and medicines. Lignosulfonates from pulping fluids are used in insecticides, cement, ceramics, fertilizers, and cosmetics. Thus, a single tree, felled by a logger who sipped his coffee in Gary Snyder's predawn dark, is reincarnated as a myriad of products whose appearances show little connection to their arboreal source.

CHEMICAL FACTORIES

Another way to understand the ways in which trees serve us is to look at the chemical compounds they contain and how these function to ward off parasites or slow decomposition or decay. For tree physiologists, primary plant compounds are those that, absorbed from the environment, contribute directly to basic metabolism, such as carbon dioxide, nitrogen, and oxygen. Secondary plant compounds, in contrast, are not part of primary metabolism. Indeed, for some time they were thought to be waste products. Biochemists now understand that they perform important functions for plants, defending them against herbivory, protecting them against disease-causing pathogens, and helping attract pollinators and fruit dispersers. They are classified into three chemical types: terpenes, phenols, and alkaloids. They are usually complex compounds and require a good deal of energy to make and maintain. Because many of these chemicals are toxic, trees must handle them by storing them within vacuoles, cavities in plant cells that are isolated from the rest of the cell by membranes.

Why do plants create these "expensive" materials? Some of these toxic secondary compounds make foliage distasteful and so deter animals such as leaf-feeding beetles or deer, whereas others interfere with the attacker's growth cycle or its ability to digest the plant. Other secondary compounds, the so-called pheromones, or "infochemicals," convey information from one organism to another. For example, in the ponderosa pine forests of Colorado, trees emit the chemical α-pinene when mountain bark beetles attack them. A sort of chemical Paul Revere that wafts through the forest, this stimulates a protective reaction in neighboring trees. By pumping more fluids into their own defense systems, nearby trees can physically expel the invading beetles with blobs of sticky resin.

The presence of secondary compounds in trees affects not only their own enemies but those of humans as well. One group of secondary compounds, the terpenes, helps your cedar deck remain intact with almost no maintenance. When it was alive, the cedar tree was protected from pathogenic fungi by those terpenes, and now, after its death, the same chemicals continue to prevent decay in the lumber milled from that tree. A subset of terpenes, called limonoids, is derived from citrus trees and from tropical trees in the family Meliaceae. This family of compounds—of which more than three hundred have been identified since 1992—is a natural insecticide. Limonoids promote a wide range of biological activities, including inhibition of the feeding mechanisms and suppression of growth of beetles, flies, caterpillars, and grasshoppers. Agricultural entomologists have tremendous interest in understanding the chemical and biological aspects of this group of compounds, as they can foresee its use as an insecticide in the hugely valuable citrus industry.

DELIVERING THE GOODS

One of my students decided to investigate how the products at the mall got to the stores in the first place. He learned that many transport mechanisms themselves involve tree products. Historically, wooden boats and wooden barges were the main means of shipping goods from one place to another. Although humans now rely on ships and airplanes made of metal for long-distance transport, trees still figure into the regional and local transportation of our commodities. In 2000, for example, well over half of the $1.7 trillion worth of goods that entered and left the United States used some form of solid-wood packing material, such as pallets and crates. In 2001, an estimated two billion pallets were in use in the United States—six for every American. Over half of these are designed to make just one trip, and pallets as a whole average just 1.7 trips. Only about 10 percent are recycled, ground up and used as landscaping mulch, animal bedding, or core material for particle board. The wood in the pallets that are discarded each year is enough to frame 300,000 average-sized houses. Each year, too, 500 million more pallets are made, consuming trees on the equivalent of 18,000 acres.

Our global reliance on pallets also introduces nonnative pests. One is the Asian long-horned beetle, an "exotic" pest that has threatened North American hardwood trees such as maple, elm, birch, poplar, and willow since 1996. The clue that these large beetles arrived in "Trojan pallets" was that outbreaks were concentrated near warehouses in New York, New Jersey, and Chicago, which contain pallets from China and Korea, where the beetles are native. Since then, infested pallets have been intercepted by vigilant entomologists in many North American cities, and so far, serious outbreaks have been contained. Europe, meanwhile, is suffering from an invasion of the pinewood nematode, thanks to products received from the United States, China, and Japan. Because of such threats, many export companies have begun to use metal or plastic pallets. These in turn create other problems, as those materials are not as easily recycled.

Moving goods from market to home entails choices in which trees play a role. Ever since polyurethane bags arrived in supermarkets in the 1980s, consumers have faced the question "Paper or plastic?" Paper bags are a forest product, manufactured from wood pulp. Plastic bags are a petroleum product, manufactured from waste products produced during oil refinement. Both require energy to make and distribute, and both have an impact on the environment during production and after disposal. Despite the more wholesome "feel" of an old-fashioned paper bag, the Institute for Lifecycle Environmental Assessment considers plastic the better choice. At current recycling rates, two plastic bags use less energy and produce less waste than a single paper bag. The energy to produce bags in the first place favors plastic over paper (594 vs. 2511 BTUs, respectively), as does the energy to recycle bags once (17 vs. 1444 BTUs). Although paper bags can be composted at home, in landfills paper does not decompose at a significantly faster rate than plastic bags. The best choice, of course, is to reuse bags, whether paper, plastic, or—best of all—fabric, thereby reducing the number of bags we consume.

TREES AT THE TABLE

> Tall thriving Trees confessed the fruitful Mold:
> The reddening Apple ripens here to Gold,
> Here the blue Fig with luscious Juice overflows,
> With deeper Red the full Pomegranate glows,
> The Branch here bends beneath the weighty Pear,
> And verdant Olives flourish round the Year.
>
> —*Homer,* The Odyssey

Some secondary compounds are attractive to pollinators. The delicate fragrances of flowers signal a nectar reward for any insect or bird that comes to visit and compel them to move from one inflorescence to another. This action results in pollination, with the resulting production of fruits and seeds. If my students and I had taken a trip to the grocery store instead of the shopping mall, we could have generated an equally long list of foods that are borne on or otherwise derived from trees. The produce section in particular reveals the bounty and diversity that trees provide for humans.

Meandering through other aisles of the grocery store reveals a profusion of tree-derived goods whose origins are less apparent. The humble pack of chewing gum occupies a strategic spot near the checkout stand, but not everyone who passes by is aware of its tree-entwined history. The ancient Greeks chewed a gummy substance named *mastiche*, derived from the resin of the mastic tree, a small evergreen belonging to the family Anacardiaceae along with mangoes, pistachios, and poison ivy. In Central America, the Mayans collected chicle, the coagulated sap of the sapodilla tree, the raw material for what has since become the worldwide chewing gum industry. These trees are concentrated most heavily in the Yucatán Peninsula. Beginning when they are twenty-five years old, they are tapped every four years, with each harvest yielding about two pounds of gum. Other trees from Latin America and Southeast Asia, as well as pine trees from the coastal states of the southeastern United States, contribute their resins to the chewing gum industry as well.

In the baking aisle, we encounter the charming deep brown bottles of vanilla. This extract comes not from a tree but from a fragrant tropical orchid, *Vanilla planifolia*, that depends on trees. A climbing perennial vine found in the forests of Central and South America, the vanilla plant grows by clasping a tree trunk with its aerial roots. It was first employed as a flavoring by pre-Columbian Mesoamerican peoples; the Aztecs used it to flavor their chocolate at the time of the arrival of Hernán Cortés. Other *Vanilla* species are grown in Tahiti and Madagascar, but their extracts are less concentrated. Vanilla extract is obtained from the seed capsules, which look like large bean pods; these are wrapped and left to "sweat" until they become black and dried. The six-week-long fermentation process causes glucosides, another type of secondary compound, to decompose into glucose and vanillin, an aromatic compound that yields the vanilla extract my daughter adds to the cookies she bakes.

The demand for natural vanilla has always exceeded the supply of vanilla beans. The chemical structure of the compound was first decoded by German scientists in the 1870s. In the 1930s, chemists managed to extract it from clove oil. By the 1940s, artificial vanillin was produced as an unlikely by-product from the breakdown of lignin in the manufacture of paper from wood pulp. This synthetic vanillin is not as high quality as its rainforest counterparts, but it is much less expensive.

Other tree-derived goods include spices, flavoring agents associated with the Far East and the Indian subcontinent. Because I have a multitude of cousins in India, several years ago when my husband and I attended a tropical biology meeting in Bangalore we took our children along and stayed on for a visit. One of the quiet highlights was a tour around the farm of my cousin Arun and his wife, Masabi. After the requisite exchange of gifts over cups of hot, sweet tea accompanied by *perus*, sweet snacks that Masabi had made, Arun took us down a narrow path to his groves of fruit and spice trees. He bent to recover a recently fallen coconut and, with a single slice of his machete, cut it open and offered us a cool drink of coconut water. He reached up and plucked a nut from a branch above his head that, when cracked open, revealed a smooth whole nutmeg surrounded by a bright red web that looked like frayed rubber bands. Arun explained that the scarlet material was mace, which they dry and grind up and which Masabi uses, in addition to nutmeg, in her delicious perus. (It is not related to the mace pepper spray that people carry to ward off dogs, which consists of oleoresin capsicum, the extract of the dried ripe fruits of hot peppers.) Arun pointed up the trunk of one of the coconut trees, where thin green vines with dark leaves were climbing: black pepper plants. Eventually he would harvest the erect fruiting bodies, which in turn, after drying, would end up being ground and sprinkled on scrambled eggs and salads worldwide.

That tour of Arun's farm gave us a mini-introduction to the world of spices. This realm intrigued the ancient Greeks and Romans as well, who spent fortunes in Arabia, then the center of the spice trade. In the sixteenth century, spices such as ginger, cumin, cloves, saffron, and cinnamon had become so highly valued for their culinary and medicinal uses that they helped spur the Age of Exploration. The quest for the precious flavorings—and for control of the spice trade shipping routes—took the British to India, the Portuguese to Brazil, the Spanish to Central America, the French to Africa, and the Dutch to Indonesia. Why did these roots, dried seeds, and chunks of bark cause such excitement? For one thing, importing rare and costly spices was a way to gain wealth and stature.

Even more important, spices improved the palatability of dull diets; stimulated salivation, which improved digestion; preserved foodstuffs from spoiling; and disguised the flavor of foods gone bad. Spices also enhanced health. Ginger, for example, was thought to improve digestion, and nutmeg to benefit the spleen.

Today, with the globalization of world markets, nearly all of these formerly exotic and expensive spices have become commonplace. Cinnamon, for instance, the inner bark of the tree *Cinnamomum verum*, is universally used as a culinary spice in pies and cakes, and it has the added advantage of being able to withstand multiple harvests. Allspice, also used in baked goods, comes from the dried fruits of the allspice tree, a tropical American evergreen. It derives its name from its bouquet, which evokes clove, nutmeg, and cinnamon. A large hardwood tree, *Bixa orellana*, provides achiote, a bright red spice used throughout Central America. The ground seeds are mixed to form a paste or powder that seasons and colors meats and vegetable dishes.

Moving from the baking aisle to the food supplement area, we encounter many more items that are derived from trees, including vitamins. Humans must acquire their vitamins by eating particular foods, since our own bodies cannot manufacture them. An alphabet soup of compounds plays an important role in maintaining the body's functions; without these compounds, scurvy, pellagra, and loss of bone density can result. Of the thirteen major vitamins, eight are derived from trees: A, B_2, B_5, B_6, B_7, B_9, C, and E. Long before the term *vitamin* was used, British sailors learned to stock their ships with limes to protect against scurvy, a disease precipitated by vitamin C deficiency that causes softening and bleeding of human tissues. Today, many of us routinely down our daily dose of vitamin C in our breakfast orange juice. Vitamin C helps counteract the harmful effects of oxidation in living tissue that occurs with aging, stress, smoking, and exposure to pollutants. It also contributes to the production of collagen, which strengthens muscles, blood vessels, and skin. The most important botanical sources of vitamin C for humans are apricots, papayas, star fruits, and grapefruits, limes, lemons, oranges, and tangerines.

Vitamin E is another antioxidant, and is found in almonds and hazelnuts. Vitamin B in its various forms comes from an array of arboreal fruits, including avocados, bananas, prunes, citrus, and nuts. Lack of this vitamin can lead to a diverse array of health problems, such as acne, dermatitis, anemia, high blood pressure, and birth defects. Its benefits are equally diverse; not only does it promote healthy skin and muscle tone, but it also enhances the immune system and promotes red cell production. Vitamin A, which improves eye and skin health as well as reproductive health, is derived from beta-carotene, extracted from apricots, peaches, mangoes, and papayas. Thus, a fruit salad made from the harvest of tropical trees does us good both in flavor and in health.

Even sweeter than these fruits—if not quite as healthy—is maple syrup. In 1663, the English chemist Robert Boyle told his colleagues in Europe, "There is in some parts of New England a kind of tree whose juice that weeps out its incision, if it is permitted slowly to exhale away the superfluous moisture, doth congeal into a sweet and saccharin substance." Early settlers in New England learned about maple sugar production from Native Americans. (*Sinzibuckwud* is the Algonquin word for maple syrup, meaning literally "drawn from wood.") The original syrup-making processes involved collecting the sap in birch-bark containers and boiling it over hot fires for days to condense the sugars into syrup.

Although improvements in the transportation of goods—including cane sugar—led to a decline in the use of maple syrup after the Civil War, "sugaring off" remains a part of our culture. One of my students, Max Bockes, grew up in northern Wisconsin. One day in class he described in joyful detail the activities that surround his family's "sugar bush" (the sugar maple forest from which sap is harvested). He told how he and his siblings stirred the big vats of sap, guessed the yield of each tree, and griddled up stacks of pancakes for the first run of the season. What seemed most significant to Max was not the syrup, but rather the renewal of old bonds as his family walked through the snow, enjoying the sweet gifts of companionship both with each other and with the maples around them.

Collection methods have evolved into more mechanized operations since the early days of tapping sap. Max's family uses plastic tubing and gas-powered pumps to directly deliver sap to the heating vessels that render the sap into syrup. Each of their maple trees produces about four quarts of sap per day, or thirty-five to seventy quarts per year. With a sap-to-syrup ratio of about 40 to 1, the annual yield per tree amounts to less than two quarts. In a good year, the trees on their 40-acre sugar bush generate about 360 gallons of clear amber syrup. The two maple species most commonly used in commercial syrup and sugar production are sugar maple and black maple. Other species can be tapped as well, such as red maple, silver maple, and even birch, but these provide a smaller volume of sap for the effort, and lower concentration of sugar in the sap. Trees grown with sufficient moisture and nutrients and with exposed crowns have higher yields than trees in dry, shaded, or infertile soil conditions. In 2005, over 47,000 tons of syrup were produced in the United States and Canada, 85 percent of which came from Canadian trees.

The cause of maple sap flow is complex, and the process has attracted the interest of tree physiologists in recent decades. In the late summer and fall, maple trees stop growing and begin storing excess starches throughout the sapwood—that is, the live outer wood of the trunk and branches. This starch remains in storage as long as the wood is colder than 40°F. When rising wood temperatures reach 40°F, temperature-sensitive enzymes in the cells change the starches to sugars—mainly sucrose—which then pass into the tree sap. As the temperature continues to increase in late spring, those enzymes stop functioning and sugar is no longer produced. But for that brief window of alternating freeze-thawing, the sugar content of the sap hovers at 3 percent, and can get as high as 10 percent of the sap volume.

Spring sap production is a relatively rare phenomenon, and it occurs only in maples and a very few other groups of trees. In sugar maple, the wood fibers are filled with gas rather than with the water that fills the fibers of neighboring species such as willow, aspen, and oak. As temperatures drop, these gases contract—unlike water, which expands. This creates space for sugar-rich sap to be sucked up by capillary forces from the roots to the branches through the tree xylem (the water-conducting apparatus of the tree). As gases in the xylem contract during the cold period, the pressure within the stem decreases, allowing the stem to absorb water. In this process, water is drawn from adjacent cells, which in turn are refilled by water absorbed from the root. As the temperature continues to drop, that water freezes between the cells. When the temperature warms, the ice melts and the ice-compressed gases expand, forcing the sugar-rich sap out of the branches and down the stem. Thus, a successful maple syrup season for Max and his family depends on there being a combination of freezing nights followed by daytime temperatures greater than 40°F. Once a string of days occurs where nighttime temperatures no longer fall below freezing, sap flow stops. When Max bores a hole into a tree and inserts the "spile," or spike, the wood fibers that serve as sap-carrying vessels are severed, and sap drops out of the tree through the spile and into the bucket that Max has hung. When it is full, he hikes it down to the "sugar bush shed," where the wood-heated evaporating vats are waiting to drive off the water and turn the thin sap into thick, beautiful, and delicious maple syrup.

In addition to the tasty toppings for our stacks of hotcakes, trees create myriad other items for our tables. Consider, for example, wine corks. One of my favorite childhood books was *The Story of Ferdinand.* The main character was a gentle bull who would rather "just sit and smell the flowers" than spar with his feisty pasture mates. Those bellicose youngsters dreamed of being in the big bullfights in Madrid, a goal unshared by Ferdinand. For me, the most fascinating part of the book was an illustration of Ferdinand lying in the shade of a cork tree. I had never seen such a tree—one that grew real wine corks that hung like cylindrical fruit from the tips of its branches. It was many years before I realized that the cork oak provides corks not from fruits but from its bark.

Cork oaks and humans have a relationship far older than Ferdinand and his tree. Over the past two thousand years, cork has been used for fishing floats, roofing, beehives, utensils, and shoe soles. Cork also makes an ideal bottle stopper, being light and compressible; when freed from

10. *Although this whimsically charming image of young Ferdinand the Bull does not reflect the real world, cork does come from oak trees. From* The Story of Ferdinand *by Munro Leaf, illustrated by Robert Lawson.*

the bottle, it returns to 85 percent of its initial volume, thanks to the forty million air-filled cells within each cubic inch of cork. The use of cork as a stopper is mentioned in texts from the distant past, including Shakespeare's *The Winter's Tale*, where a ship plowing into the waves is likened to a cork thrust "into a hogshead." However, it was not until the advent of the uniformly shaped glass bottle in the early 1800s that corks became commonplace as bottle stoppers.

The cork oak has evolved a remarkably spongy bark. For trees, bark functions in the same way our skin does for us—to shield the delicate inner organs from the impacts of the world. Like the epidermis of human skin, which consists of multiple strata, tree bark has different layers that protect it from physical injury, fire, and temperature fluctuations. In the cork oak, in contrast to most other trees (the cinnamon tree being a notable exception), the bark can be removed without harming the tree's living parts because the outer spongy bark separates not along the vascular cambium (the living, inner layer of bark) but instead along the cork cambium (also called the "phellogen"), the outermost layer that is already dead. Once the outer cork layer is removed, the inner bark grows a new layer of cork. Cork oak forests grow in the northern regions of Spain and Portugal, though the tree is also native to France, Italy, Morocco, Tunisia, and Algeria. The tree has an average lifespan of 250 years, but some individuals have reached 450 years of age. Although a typical cork oak produces enough cork for four hundred bottles a year, the Whistler Tree in the Alentejo region of Portugal, the world's largest cork oak, annually yields over one ton of raw cork per harvest, enough for ten thousand wine bottles.

For the cork orchardist who makes a living from these trees, capital returns are slow to materialize. The first stripping of bark, called virgin cork, occurs when the tree is twenty-five years old. Because of the rapid growth of the young tree, this bark has deep furrows, and it is not suited for wine corks. Not until the third stripping, when the trees are forty years old, can the bark be used for corks. Subsequent harvests occur every decade for another two hundred years, with an average of fifteen harvests over the lifespan of a tree. Large white numbers are painted on the trunks to keep track of the harvest years. Laws mandate how frequently and at what time of year the bark can be taken from the trees; if an old tree is cut down, a young one must be planted in its place.

Skilled migrant workers harvest most of the cork in Portugal and Spain, gathering as much as 1,400 pounds each day. With small sharp axes, they cut two horizontal slices in the bark eight feet apart, then cut a connection between the two, after which they carefully peel the tree. Although in the year following a harvest new branches produce 20 percent fewer leaves, the removal of bark appears to do no long-term damage to the tree. The harvested bark is left to weather outside for six to twelve months. The weathering oxidizes the bark, allowing some of the tannins and mineral salts to leach out. The bark is then shipped to cork factories, where it is placed in boiling water to flatten and fully expand the cork cells. After being sorted for quality, the bark is cut into strips, from which corks are punched. The growth rings in the bark are oriented so that the lenticels (small openings in the bark that allow air to pass into the cambium) do

11. Cork is removed multiple times in a cork oak's life, with workers pulling the bark away from its living cambium. Courtesy Cork Supply USA.

not run the length of the cork, which would lead to leaks. The strip cutting and cork punching are done by hand, though machines are now used to cut lower-grade corks. The corks are then washed in a chlorine solution to disinfect them and lighten their color, after which they are sorted and printed with logos for individual wineries and thence make their way to candle-lit dinners in our dining rooms and restaurants.

Recently, the cork industry has been looking for ways to make these centuries-old practices more efficient. For example, the waste from cutting out the high-quality stoppers is now ground up and glued back together in the shape of a cork, creating "agglomerated" corks. However, many wineries have started using completely synthetic corks derived from petroleum products, to reduce both costs and the risk of fungal contaminants in the wine. Synthetic corks now make up 8 to 10 percent of the thirteen billion corks manufactured each year, and that proportion is growing steadily. The advent of synthetic corks will likely have a negative impact on the oak forest habitat, which sustains many species of birds, insects, and animals, including the endangered Iberian lynx and Iberian eagle, and provides wintering grounds for Europe's crane population. These forests also provide income for humans, both those who work in the cork industry and the herders of sheep and goats that graze beneath the oaks. If synthetic corks replace natural corks, the demand for cork from these trees will plummet, and owners may well replace native oak forests with pine and eucalyptus plantations to produce wood pulp for paper. Conservation groups are educating consumers to demand real cork over synthetic cork, and supermarkets are being pressured to label the type of cork being used so that customers can make an informed choice. It is sobering to think that a diner's seemingly simple request for a glass of wine can affect arboreal and human communities half a globe away.

Chopsticks are another tree-derived food accessory, one that originated in China five thousand years ago. There, food was chopped small, and a pair of sticks became the staple utensils, called *kuai-zi* ("quick little fellows"). By A.D. 500, use of chopsticks had spread to Vietnam, Korea, and Japan. In Japan, chopsticks were originally used only for religious ceremonies and were made of jade, gold, brass, coral, lacquered wood, and ivory. As their use spread into the kitchen and everyday meals, manufacturers turned to bamboo, cedar, pine, and teak. In the 1870s, the Japanese began making disposable chopsticks (*wari-bashi*, meaning "little quick ones") from the scraps left by woodworkers. Nowadays, the number of chopsticks used in Japan each year is huge—over twenty billion pairs. Because Japanese forests are protected by environmental legislation that severely limits cutting, 90 percent of chopstick wood comes from elsewhere, especially the hardwood forests of Vietnam, Malaysia, Thailand, Canada, and the United States.

China, too, uses disposable wooden chopsticks—some 450 billion pairs each year, the equivalent of twenty-five million trees. Yet China is facing international political pressure to improve its environment and cut its energy use by a fifth by 2012. In response, the Chinese government is imposing new or higher taxes on a range of goods and fuels, including *wari-bashi*, which now carry a 5 percent tax. This seems a meaningful way for the Chinese leadership to demonstrate its commitment to the environment.

TREES AND TOILETRIES

During a recent flight cross-country, I was randomly plucked from a line of waiting passengers to undergo an extra security check. As two uniformed security workers went through my tote bag of toiletries, I dealt with my irritation by describing the arboreal origins of various pertinent items—starting with the security personnel's eyeglass frames, which were made from cellulose wood fibers. When they opened my makeup kit, I pointed out that cellulose gave my lipstick its smooth texture and that my nail polish got its gloss from nitrocellulose. Tree gum, an exudate from trees belonging to the pea family, Fabaceae, made the adhesive on the bandage strips I'd stashed in my first aid kit, while methylcellulose thickened my shampoo and conditioner.

I concluded my trees-and-toiletries lecture by summarizing the long history of tree influences on toothpaste. Many ancient and current cultures, I said, use twigs for teeth cleaning and have concocted various forms of tooth powder containing abrasives and mouth-freshening herbs and minerals to help maintain oral hygiene. In India, storytellers recount the legend of the Hindu prince Siddhartha, who became the Buddha. His discarded toothbrush-twig grew into an enormous tree. Many Indians today use the same type of twig—from the neem tree—to whiten and cleanse their teeth. Muslims, too, use tree twigs to clean their teeth, heeding the advice of the Prophet Muhammad. By Islamic prescription, these twigs may be plucked from one of several trees, including the peelu tree, the olive, and the walnut. Many dental remedies are derived from trees as well, including witch hazel, used by Native Americans to reduce swelling and control bleeding, and flower buds from the clove tree, used by Africans as a remedy for toothache. A common ingredient of natural toothpastes is propolis, a sticky resin that seeps from the buds and bark of conifer trees and is harvested by bees. They carry it to their hive, blend it with wax secreted by glands in their abdomen, and apply it to the interior of their brood cells. This lining has antibacterial properties, ensuring a healthier environment for the rearing of the bees' young. In toothpastes, propolis is joined by other flavorings, such as grapefruit seed extract, cinnamon oil, and clove oil. "All of which," one of the security officers said wearily, "no doubt came from trees," as he waved me through to the checkpoint exit.

FRAGRANCES

Perfumes (from the Latin *per fumum*, "through smoke") originated in ancient cultures and were considered essential elements of civilized life, serving both religious and practical functions. In ritual, fragrant woods burned as incense carried an ethereal message of appreciation to the gods. According to the Gospel of Matthew, frankincense and myrrh were carried by the three wise men across the desert to welcome the newborn Jesus. These resins, extracted from trees in the family Burseraceae, have become synonymous with cherished religious offerings. Today, they are used in perfumes and as flavorings in beverages and toothpastes. In ancient times, too, tree resins, many of which are antibacterial, were used as preservatives and fumigants as well as aphrodisiacs. In the Old Testament, romantic verses describe the sensory, and sensual, excitement offered by various saps, spices, and extracts: "I have perfumed my bed with myrrh, aloes, and cinnamon. Come let us take our fill of love till morning" (Proverbs 7:17–18). The Hindus' *Kama Sutra*, too, extolled the use of nutmeg, cloves, cardamom, and ginger for greater fulfillment in physical love. Today, these materials are included in massage oils.

Sandalwood is one of the oldest perfumes, with origins in India. Its oil is extracted from the wood and roots of *Santalum album*, and it has a sweet, woody scent. It is my favorite perfume, and I hoard the small orange bottles of sandalwood paste that my Indian relatives send me from across the ocean. Rosewood oil, a fragrance in soaps, is extracted from the trunks of tropical trees of the genus *Aniba*. The demand for rosewood oil in perfumes rocketed in 1921, with the introduction of the popular fragrance Chanel No. 5. Destructive extraction methods used to obtain the oil soon led to the decimation of the *Aniba* tree in the Amazon. In the 1980s, the Brazilian government designated it an endangered species. This forced the rosewood oil trade onto the black market, which made it nearly impossible to regulate and drove prices up. The use of cheaper synthetic alternatives and an innovative branch-harvesting operation have since met the continuing demand.

TURPENTINE

Although turpentine itself is not a denizen of makeup bags or bathroom cabinets, materials derived from turpentine are used in perfumery. Turpentine is tapped from pine trees, as I learned when I married into the Longino family. My father-in-law, B. T. Longino, raises beef cattle on a vast flat stretch of southern Florida. From the porch of the home where he and his wife, Jane, live, he looks out over wide grass pastures, with herdlets of mother cows and their calves, an occasional pond, and oak-palm hammocks—islands of oak and palm trees—dotting the landscape. Just fifty years ago, this ranch was an unbroken forest of pine trees. B. T.'s father and his work crews harvested resin, tapping the slash pines in the same way that the *chicleros* harvest rubber in South America. Workers would cut deep, V-shaped incisions into the bark, and the sap flowed out into small bowls they attached and later collected. They then boiled batches of the resin in cauldrons, distilling the resin into its various parts. Today, when our family takes rambling drives around the ranch in B. T.'s ancient swamp buggy, we may spot an old pine tree with a "cat-face," the scar of those V-shaped slashes that were cut into the cambium of the trees. Occasionally on our walks, we come upon a pottery shard or, rarely, an intact ceramic bowl that held the pine sap decades ago. It makes the ghosts in the pinewoods come to life, and the history hidden in the artifacts more real.

For several decades, the sap from these Florida pinewoods yielded "naval stores," resin-based products used in ship construction. Turpentine, one of the distillates, is composed of terpenes—volatile organic compounds found in many trees, mainly conifers. It historically has been used in diverse ways, including as a medicine against lice and as an internal treatment for parasites. In the industrial world, it functioned as a thinner of paints and a component of varnishes. Along with another tree, the camphor laurel, it currently provides the raw material for producing camphor, which has such diverse uses as a moth repellent, constituent of embalming fluid, and fireworks ingredient. Another turpentine derivative, rosin, is used to increase the friction on the bows of stringed instruments so they produce a more beautiful tone. Rosin is also used by ballplayers to get a better grip on balls and bats, and by ballet dancers to give the toes of their shoes more grip on the dance floor. Today, however, humans are rarely involved in the extraction of turpentine from live pine trees. Nearly all naval stores are now obtained from stumps and waste wood (through steaming or chemical extraction), by-products from the paper-making process, and petroleum. Nevertheless, while the pines on the Longino Ranch no longer provide a harvest of sap, they still give a lifetime of shade to the livestock and a sense of tall, spare beauty to the land.

COMMUNICATION

What would humans do—or be—without paper? The books and newspapers we read each day, the shopping lists we scribble, and the letters we receive have not yet been wholly replaced by electrons on computer screens. Every year, more than two billion books, 350 million magazines, and 24 billion newspapers are printed. Two computer scientists, Roy Rada and Richard Forsyth, in their book *Machine Learning*, go further than most writers in acknowledging their debt to their sources: "Many people, other than the authors, contribute to the making of a book, from the first person who had the bright idea of alphabetic writing through the inventor of movable type to the lumberjacks who felled the trees that were pulped for its printing. It is not customary to acknowledge the trees themselves, though their commitment is total."

How much paper *do* we use? According to figures published by the Food and Agriculture Administration in 2007, each year Americans use more than 90 million tons of paper and paperboard, or 950 pounds of paper products per person—one-third of the world's paper. People in developing countries, in contrast, use only 35 pounds of paper a year on average. In India, the figure is 10 pounds, while in twenty countries in Africa it is less than 2 pounds. Although the advent of the computer age paralleled a reduction in the rate of increase in paper consumption in developed countries such as the United States and the United Kingdom, global paper use increased more than sixfold over the second half of the twentieth century, and has doubled since the mid-1970s.

Producing one ton of paper requires two to three times that weight in trees. In the papermaking process, wood is first chipped into small pieces. Then water, heat, and sometimes chemicals are added to separate the wood into individual fibers, which are mixed with more water and recycled fiber. This pulp slurry is sprayed onto flat wire screens that are moved through the drying machine, where water drains out and the paper fibers bond together. The web of paper is pressed between rolls to squeeze out yet more water. Heated rollers then dry the paper, and the paper is slit into sheets. Newly cut trees account for 55 percent of the global paper supply, while recycled wood-based paper and cardboard make up 38 percent. Non-tree sources—such as crop waste from rice, wheat, and cotton, as well as woody and herbaceous plants such as sisal, hemp, kenaf, and agave—constitute the remaining 7 percent.

Many of our writing tools are also derived from trees. In ancient Rome, scribes wrote on papyrus with a thin metal rod called a stylus. Some styluses were made of lead, which is why today we call the core of a pencil the "lead" even though it is made from graphite, a hard, stable form of carbon. Graphite leaves a darker mark than lead but is so brittle that it requires a holder. At first, sticks of graphite were wrapped in string, but they were later inserted into wooden sticks hollowed out by hand. The first mass-produced pencils were made in Germany in 1662 and were left unpainted to show off their high-quality wood casings. By the 1890s, however, many manufacturers were painting brand names on them.

Have you ever wondered why pencils tend to be yellow? During the 1800s, the best graphite in the world came from China. American pencil makers wanted to inform customers that their pencils contained Chinese graphite. By painting their pencils bright yellow, they communicated a connection with China, where yellow is associated with royalty and respect. Today, 75 percent of the pencils sold in the United States are still painted yellow. Early American pencils were made from eastern red cedar, a strong, splinter-resistant wood that grows in the southeastern United States. By the 1900s, pencil manufacturers had turned to California's forests of incense-cedar, which is now the wood of choice for domestic and international pencil makers.

Although the most obvious tree-centered modes of human communication are paper and pencil, another form is the totem pole used by Native American tribes of the Pacific Northwest coast. Most totem poles are carved from western red cedar. Their original makers included the Haida, Tlingit, and Tsimshian peoples, but other groups have assumed the tradition as well. Totem poles announce family/clan status or history, mark memorial sites, and relate significant events or the stories of important people in the clan's history. Although some carved symbols or characters may be identifiable and common among different poles, each totem pole has its own unique history and meaning.

AT THE END OF LIFE

Trees and wood have been associated with death for a very long time—not just in socially acceptable ways, in the form of coffins and grave markers, for example, but in less savory ways as well, as instruments of capital punishment and murder by hanging, lynching, and crucifixion. The origins of hanging are obscure, but suspension from trees and later scaffolds or gallows had become one of the most common forms of execution in Britain by the tenth century. Gallows evolved from a simple cross-beam construction to multiple-hanging scaffolds. Hanging is growing less common, but it continues to serve as an acceptable form of capital punishment. In twentieth-century America, illegal forms of hanging in which victims were suspended from trees by mobs of civilians were often spurred by racial hatred.

Although the most famous crucifixion, that of Jesus Christ, was performed on a wooden cross, crucifixions originally were performed on a single wooden post or tree trunk with the victim's arms secured above or behind. The Old Testament (Deuteronomy 21:22–23) had specific directives on the procedure: "And if a man have committed a sin worthy of death, and he be to be put to death, and thou hang him on a tree: His body shall not remain all night upon the tree, but thou shalt in any wise bury him that day; (for he that is hanged is accursed of God;) that thy land be not defiled, which the Lord thy God giveth thee for an inheritance." Dogwood was said to be the wood used in the cross on which Christ was crucified. There is a myth that this tree was ashamed of the role it played in history, and in compassion Jesus had it reduced from a tall, sturdy, oak-like tree to the small, skinny-limbed tree we know today. Because of its fragile form, it would never again be used for making a cross.

Trees are also used for more peaceful aspects of death. According to the Casket and Funeral Supply Association, in 2007 about 73 percent of the total 2,380,926 deaths required use of a traditional casket. Of these 1.7 million caskets, over 300,000 (15 percent) were made of wood, mainly poplar, oak, and pine. Surprisingly, the great majority of caskets are made of steel, while a few percent are made of plastic or fiberglass. Although traditionally the "plain pine coffin" is a symbol of frugality and simplicity, nowadays we have the even more ecological option of green burial. At Memorial Ecosystems in South Carolina, for example, the traditional elements of a cemetery—manicured lawns, gravestones, and metal vaults—are nowhere in evidence. The deceased are buried in caskets made from biodegradable wood, and no toxic embalming fluids, which can seep into groundwater, are permitted. Rather than vaults and headstones, white oaks and other trees serve as memorial markers, while wildflowers carpet many graves. The grounds, a 32-acre nature preserve, are managed as a native ecosystem. Dr. Billy Campbell, who founded Memorial Ecosystems in 1998, got his idea from Borneo, where "spirit forests" are maintained, leaving the dead undisturbed in sacred ground. The Centre for Natural Burial monitors the growing interest in this way of leaving the earth with minimal impact, and even offers a quarterly newsletter. In 2007, this organization listed six green burial sites in the United States, with four others pending, so getting an ecological send-off is becoming easier.

In some cultures, wood is used not to preserve the body but to dispatch it into the afterlife. The Vikings often burned their dead leaders in ship burials. A Viking was said to achieve immortality if his corpse was incinerated on a great seafaring warship. Although pyres—mounds of wood for burning a body—are no longer used in Western practices of death, in Hindu tradition they are common. Hindus believe in reincarnation, the idea that after death a person will be born again but in another form. A person's future form depends on his or her behavior in the past, with good actions resulting in reincarnation as a higher form or caste and bad behavior leading to rebirth as, perhaps, a dog or a worm. Cremation of the body is the act that releases an individual from the past life into whatever form the next life will be. During the ritual, the body is laid out on blocks of wood. The eldest son circles the body with a flame three times, and then performs the difficult act of lighting an oiled cotton wick placed in his parent's mouth, which ignites the pyre. The travel blog of a young woman, Romena Romuar, who witnessed the funeral pyres on the cremation grounds of Pashupatinath, a Hindu temple in India, describes the scene and its attendant emotions well:

> Brilliantly red and orange flames devoured all things in its path, crackling and popping as it turned wood into embers and flesh into ashes. The scent of cinder wood barely masked the unforgettable odor of burning flesh. Smoke continued to billow through the evening sky and against the veil of darkness,

resembling ghosts of those leaving the material world, serving as tangible evidence that their souls, indeed, ascended directly up to the heavens. The ritual reminded me that we leave the world with the very same things we were born with—our bodies, our hearts and our minds. Everything else is left in the material world. Everything else doesn't matter.

ENRICHING THE MATERIAL WORLD

What does he plant who plants a tree?
  He plants a friend of sun and sky;
He plants the flag of breezes free;
The shaft of beauty, towering high.
He plants a home to heaven anigh
For song and mother-croon of bird
In hushed and happy twilight heard—
The treble of heaven's harmony—
These things he plants who plants a tree.

What does he plant who plants a tree?
  He plants cool shade and tender rain,
And seed and bud of days to be,
And years that fade and flush again;
He plants the glory of the plain;
He plants the forest's heritage;
The harvest of a coming age;
The joy that unborn eyes shall see—
These things he plants who plants a tree.

What does he plant who plants a tree?
  He plants, in sap and leaf and wood,
In love of home and loyalty
And far-cast thought of civic good—
His blessing on the neighborhood
Who in the hollow of His hand
Holds all the growth of all our land—
A nation's growth from sea to sea
Stirs in his heart who plants a tree.

—Henry Cuyler Bunner, "The Heart of the Tree"

The goods and services that trees provide pervade all branches of our economic system. In 2007, the Parks Department of New York City attempted to calculate the monetary value of the city's trees. The first step, which took more than a thousand volunteers two years to perform, was to carry out a census of all the trees on city streets. They arrived at a figure of 592,130, which is in addition to the roughly 4.5 million trees in parks and private gardens. The data were fed into a computer program, called Stratum, which was developed by researchers at the University of California, Davis, and in the U.S. Forest Service. The program estimates such factors as a tree's impact on property values, the amount of carbon dioxide and other pollutants removed from the air by trees, and the amount of energy conserved by shading and transpiration (the movement of water out of leaves during the process of photosynthesis, which can significantly cool surrounding areas). Their conclusions: New York City's street trees provide an annual benefit of about $122 million, equivalent to each New Yorker receiving $5.60 in benefits for every dollar spent on trees. This information gave the Parks Department hard data for budget discussions—a persuasive complement to aesthetic, recreational, and spiritual arguments for planting more trees.

That initial list of products generated by my students continued to grow throughout the semester, along with our awareness of the sources of the things we use. Even now, six years later, my students still send me sporadic messages about objects we missed: boats (in the sea, with wooden figureheads, and in glass bottles, in miniature), musical instruments (mandolins and flutes), airplanes (Howard Hughes's *Spruce Goose*), coffee stirrers, fence pickets, and rayon. Each of these objects contains a tree story that extends into our human stories—serving as arboreal bookmarks in our existence.

Chapter Three

SHELTER AND PROTECTION

How sweet to be thus nestling deep in boughs,
Upon an ashen stoven pillowing me;
Faintly are heard the ploughmen at their ploughs,
But not an eye can find its way to see.
The sunbeams scarce molest me with a smile,
So thick the leafy armies gather round;
And where they do, the breeze blows cool the while,
Their leafy shadows dancing on the ground.

—*John Clare, "In Hilly-wood"*

When I was nine years old, my father designed and built a magnificent tree house in the shape of a boat. It nestled in the airspace between two giant linden trees in our backyard. During its construction, my siblings and I would run home from school to watch the progress being made sixty feet above our heads. Even though its dimensions were only ten by twelve feet, the boat seemed not a cubit smaller than Noah's Ark. When it was completed, we had the most wonderful place to spend the night, complete with a wooden figurehead in the intended shape of an ancient turtle, a trapdoor, and a pulley system to haul up snacks, sleeping bags, and flashlights. For that evening until the next morning, the world was ours for as far as we could see. The structure reminded my siblings of the *House at Pooh Corner* chapter "In Which Piglet Is Entirely Surrounded by Water," when the timid Piglet holes up in his tree during a long rainstorm and waits, as the water steadily rises, to be rescued. For me, our treetop shelter was in turn a rescue vessel in case of neighborhood flooding, a sanctuary for injured birds, a refuge for wounded soldiers, and a safe haven for Anne Frank. It was our ark.

Something about being in a wooden tree house fosters in me—and in all but the most acrophobic—a feeling of safety. Treetop refuges exemplify the ways that trees fulfill the second level of the modified Maslow pyramid, the need for shelter and security. Trees provide not only housing but also fuel for warmth, shade against the heat, private spots in public places, and aesthetic beauty. In addition, they give us a spiritual sort of security, one that I cannot describe but that is as real as a stack of two-by-fours.

TREETOP AND TERRESTRIAL SHELTERS

Peter Nelson, a tree house builder and author, wrote: "Tree houses inspire dreams. They represent freedom: from adults or adulthood, from duties and responsibilities, from an earthbound perspective. If we can't fly with the birds, at least we can nest with them." His books document a variety of tree houses, from casual shacks made from discarded lumber to multitiered minimansions paneled with cedar and entered through french doors. Browse the carpentry section in a bookstore and you will find how-to manuals for building tree forts, tree houses, and tree platforms. Some of the plans are structurally elaborate, requiring knowledge of complex joinery and wind stresses on multiple trunks. Nearly all of them emphasize the need for care in fastening these structures onto trees—not in the sense of structural integrity, but for safeguarding the tree itself. Placing bolts into the cambium can introduce disease, and constrictions around the girth of a trunk or branch must allow for movement and growth of the tree.

Tree house enterprises have greatly expanded since the days of my father's backyard construction. You can now make reservations at tree house hotels and resorts all over the world. In 2005, my husband and I took sabbatical leave and our family embarked on an overland journey across the United States. Our first night's stop was the Out 'n' About Treesort, in Takilma, Oregon, snuggled in a valley of the Siskiyou Mountains. This licensed bed-and-breakfast, established in 1996, has a twist: all of the guest units are perched in the treetops among the branches of a conifer grove. The Treesort is booked months in advance by people seeking a balance of repose and adventure. The owner, Michael Garnier, a wiry and intense man with strong hands and forearms, possesses a quiet power that matches the passion he had to create this place. "It took eight years of legal battles with Josephine County officials before they recognized the soundness of my designs and gave me permits to be legal," said Mike. "We were ordered to shut down our tree house rentals several times, and were actually ordered to tear down the tree houses at one point." His zeal for trees is reflected in the signs that are scattered through the compound, marking the Treezebo, the Treepee, the Serendipitree.

My family stayed in the Forestree, which had a tiny woodstove, bunk beds, a wee table, and a small bathroom, all sixty feet above the ground and linked to the next unit over by a wonderfully jiggly suspension bridge. After lights out, we shared a long silence as we adjusted to sleeping suspended in the air, and then assured each other that "that swaying feeling" would no doubt enhance our rest. After what did turn out to be a remarkably peaceful night aloft, Mike took the morning to show us all eighteen tree houses. I was impressed with how he had adapted each structure to the unique architecture of its supporting tree. Trees and human structures blended nearly seamlessly, demonstrating how—with thought, experience, and respect—humans can live harmoniously with nature.

Trees have provided humans with shelter for over one hundred centuries. Evidence of timber construction in some of the earliest civilizations can be found in Turkey, Mesopotamia, Israel, and Lebanon. Remnants of burnt wood and postholes reveal that numerous tree species were used, including cedar, pine, juniper, oak, spruce, cypress, walnut, maple, and ash. The cedars of Lebanon described in the Bible had trunks as large

12. Recreational tree climbing reaches comfortable and intriguing heights at southwest Oregon's "Treesort." Courtesy www.treehouses.com.

as thirty feet in circumference, with wide-spreading branches clothed in bright green needles. The Bible mentions the welcome shade they provided, and their use in the construction of temples and palaces. In 1 Kings (6:9–10, 7:2) we read: "So he built the house, and finished it; and covered the house with beams and boards of cedar. And then he built chambers, five cubits high: and they rested on the house with timber of cedar. . . . He built also the house of the forest of Lebanon upon four rows of cedar pillars, with cedar beams upon the pillars."

On the other side of the globe, in ancient China, all parts of a wooden building were joined by mortise-and-tenon joints, the pieces fitting

13. Gustave Doré's Cutting Down Cedars for the Construction of the Temple *depicts a historical use of the famous cedars of Lebanon, which were overharvested centuries ago for their valuable timber. The stands have never recovered their former extent.*

together like a puzzle, without metal nails or screws. In Japan, timber construction began in simple forms, some of which embodied the Shinto belief that trees are homes of the spirits and so need a central place in the homes of people. The Yayoi dwellings of ancient Japan (ca. 300 B.C.–A.D. 300) consisted of a conical timber-and-thatch cap erected on a central pillar over an earthen dugout. This basic structure has remained an element of Japanese architecture, both for its spiritual significance and for its ability to withstand earthquakes.

In North America, the longhouse was common to the Iroquois and other tribes of the Great Lakes region, as well as to the natives of the Pacific Northwest. A longhouse sheltered multiple families, all of whom shared a single fire. Its construction consisted of rows of saplings driven into the ground that were bent inward to form the roof. Bark and additional supports provided insulation, and gaps were left in the roof to vent smoke from the central fire. The wigwams of the Algonquin peoples were also made of tree saplings, which, arranged in a circle or oblong, were bound together at the top in a rounded shape and covered with skins or bark. In contrast, the tipis of the North American Plains peoples were made of wooden poles arranged conically and covered with bison hides; they were easily dismantled, which complemented these peoples' nomadic way of life. Similar to structures of other tribes that were tightly linked to land and forest, tipis served both a protective and a spiritual purpose, with the earthen floor linking the inhabitants to the ground and the wooden poles connecting them to the sky. North American coastal peoples with plentiful forests often created cedar plank houses using post-and-beam construction.

Sweat lodges were another type of construction common in the Americas, from Alaska to the Yucatán. Mikkel Aaland describes a typical structure:

> Sweat lodges were usually domed. Nomadic tribes drove pliant boughs, such as willow, into the ground and arched them into a hemisphere, secured with withes [twigs]. Stationary tribes used more substantial materials—logs and heavy bark. A depression was dug near the door or in the center to cradle the

> rocks, which were heated outside and brought in on forked sticks. Steam was produced by sprinkling the rocks from a hollowed buffalo horn. . . . After hours of talk, gossip, and dancing the fire was fed to a noble size; the lodge became torrid and sweating began.

To protect themselves from the caustic smoke of the sweat lodge, which had no outlet, participants used fine shavings of willow or spruce to create oval-shaped respirators that covered the lower face. The value of the wooden sweat lodge was apparent to Roger Williams of Rhode Island, who, writing in 1643, said of the local natives: "They use sweating for two ends: first to cleanse their skin; secondly to purge their bodies, which doubtless is a great means of preserving them, especially from the French disease [probably influenza] which by sweating and some potions, they perfectly and speedily cure."

Immigrants from western Europe brought with them to America their own traditions of log construction. The people of Scandinavia, Germany, and northern Russia, especially, had great experience in building with softwoods, which are easily worked with hand tools. As new settlers made their way across the continent, they took advantage of the forests they encountered. Small groups of people could fell, trim, notch, and transport enough trees to put up a small cabin in a matter of days. To build a typical log cabin of 400 square feet (twenty feet wide by twenty feet long by eight feet tall), the work party would have to fall and buck twelve trees that were thirty inches in diameter and eighty feet long. Today's houses, though more complex, do not necessarily use more trees. A typical "stick-built" wooden frame house of 2,000 square feet, with one story and a basement, requires roughly 15,800 board feet of lumber for the structure and the framing and another 10,000 board feet for other purposes (a board foot being a piece of wood twelve inches wide, twelve inches long, and one inch thick). Only three trees of the same dimensions as were used for the log cabin would provide the timber for a place we now call home.

AT ENVIRONMENTAL EDGES

Both humans and trees are capable of adapting to even the harshest environments. In the summer of 1975 I flew over the black spruce forests—the taiga—of Alaska, on my way to pan for gold in the Big Hurrah country north of Nome. From the window of the little plane I could look down on a vast expanse of trees that would soon face subzero temperatures and endless night. When I arrived on the tundra, I was equally amazed at the wooden structures built by people who had panned for gold the century before. Although those people and the gold are long gone, their houses remain, evidence of a hard-won existence amid the darkness, isolation, and cold of the far north. How do trees and humans survive? How have their functions and behaviors evolved under such extremes?

High-latitude and high-altitude trees must adapt to cold weather, frozen soil, and extreme exposure to winds. In tundra environments, roots are restricted to the upper layer of soil that thaws during the six-week growing season. Growth tends to be stunted and slow, assuming a form called krummholz (German for "twisted wood"). Krummholz is typical also in alpine zones, where trees such as birch and willow, rarely growing taller than three feet, become prostrate, looking for all the world as if they are crawling along the ground in very slow motion. This growth habit allows them to withstand the effects of tons of snow and wind-driven ice without breaking.

These humped, shrublike trees create symbiotic relationships with local fauna. The cavities within their branches provide refuge for animals such as marmots and chickarees, chipmunklike rodents that skitter about the tundra surface in the summer and retire to underground nests during the winter. These animals in turn contribute organic matter for the tree from the decomposition of their nesting material and feces, and in the course of moving from one tree cave to another, they also move cones and seeds around the entire habitat. Thus, by providing shelter to their mobile cohabitants, the dwarfed trees enhance their own nutrient sources and ensure the dispersal of their own seeds.

14. The prostrate, shrubby structure of trees at alpine and arctic tree line enables these species to survive extreme physical exposure and very short growing seasons. Photo by Greg and Mary Beth Dimijian.

Just as trees at tree line hunker close to the ground, where wind is slowed due to friction, human shelters in arctic and alpine habitats are traditionally low to the ground and rounded. The humped architecture allows air to pass over the structure rather than penetrating inside. In the far north, an igloo (an Inuit word meaning "dwelling") is created from blocks of snow. As they spiral upward, the igloo walls curve in, creating an ice dome with a hole at the top for ventilation. Indigenous people who live a bit farther south build structures of a similar shape, subterranean dugouts capped with driftwood.

The challenge for trees living at the other extreme, in very hot, dry places, is obtaining enough water under conditions of desiccating sun and low, unpredictable rainfall. How do desert trees cope? One technique that they have evolved is to reduce water loss through their leaves by doing a "switch play" with their plant parts. For example, the elephant trees of the

15. The broad-crowned drought-tolerant umbrella acacia appears throughout dry areas in Africa. Photo by Greg and Mary Beth Dimijian.

Sonoran Desert drop their minute leaves during drought conditions. The green bark that underlies the thin outer papery bark steps in to do the leaves' normal work of absorbing sunlight and transforming it into sugar.

Other desert trees, called phreatophytes (from the Greek *phreat-*, meaning "well")—among them the acacias of Africa and the mesquites of Central America—maintain their moisture balance by sinking roots deep into the ground, where they can tap into the permanent water table. Getting established is a huge challenge for these plants, as the taproot must travel downward as far as fifteen stories through bone-dry soil. Therefore, phreatophytes tend to be restricted to places such as dry riverbeds where the soil is occasionally wetted.

A third way trees survive in the desert is by capturing water during short wet seasons or sporadic rainstorms and storing it in their swollen trunks. In the dry deciduous forests of Africa, the baobab tree, which serves as a landmark for travelers, a gathering point for villagers, and a refuge for animals, uses this strategy. Its gigantic trunk, which is often wider than it is high, functions as a huge moisture storage organ, having a spongy and absorbent texture and, as it grows older, a hollow interior, which be-

16. These desert-dwelling baobab trees serve as community gathering spots, in part because of their remarkable ability to retain water, which can be tapped for drinking during long dry periods. Photo by Greg and Mary Beth Dimijian.

comes a literal water tank. Tribesmen in the Kalahari gain access to these reservoirs—which can hold more than 30,000 gallons—by sipping through grass straws. The baobab's branches are devoid of leaves for nine months of the year, which minimizes water loss while making the crown look like an upside-down root system (hence its nickname of "upside-down tree"). Euphorbia trees of Africa also have corrugated trunks that store water, swelling up like a bellows after a rainstorm. Although euphorbia trees are similar to the baobab in structure and function, the two are completely unrelated biologically; their parallel adaptations occurred by convergent evolution.

Unlike trees, of course, humans are mobile. Instead of going dormant or storing water in our bodies, we tend to cope with dry, hot environments by moving to available water. Many desert-dwelling cultures are nomadic, and they create structures that include wood. Similar to the tipis of the Plains Indians, for example, the homes of nomadic herders of Somalia are portable dome-shaped structures, called *aqal.* These also converge in form with the yurt, or *ger,* the traditional dwelling of Mongolians, which is squat, rounded, and designed to withstand the extremes of the Gobi Desert. The structure can collapse to fit on one draft animal and can be set up again in half an hour. The walls are a criss-crossed lattice constructed of wooden poles joined with leather lacing; in the roof is a hoop of wood with slots into which the roof poles can lock and upon which the felt is draped.

The Bedouins of northern Africa, who migrate into the desert during the rainy season and return to cultivated areas during the dry season, have a similar tradition. Their living structure is known as the *beit al-sha'r,* or "house of hair," after the coverings, which are woven from the wool of their sheep and goats and supported by wooden poles. The shelter provides shade from the hot sun and insulation on cold desert nights. The flattened roof keeps the desert tents from blowing away, showing a kinship not only with arctic igloos but also with the compressed and flattened krummholz of mountaintops.

WARMTH

When wintry days are dark and drear
  And all the forest ways grow still,
When gray snow-laden clouds appear
  Along the bleak horizon hill,

When cattle are all snugly penned
And sheep go huddling close together,
When steady streams of smoke ascend
From farm-house chimneys,—in such weather
Give me old Carolina's own,
A great log house, a great hearthstone,
A cheering pipe of cob or briar,
And a red, leaping light'ood fire.

—*John Henry Boner, "The Light'ood Fire"*

My family and I live in the Pacific Northwest, which is neither terribly cold nor horribly hot. But our winters are wet, dark, and very long. When the globe turns us toward the winter solstice and daylight lasts less than seven hours, our home life becomes increasingly oriented around the woodstove that stands in one corner of our living room.

We gather our fuel from the forested acres that surround our house—alder snags, maple boughs, and fallen Douglas-fir trunks. My husband, who calls his chainsaw "Little Shiva" after the Hindu god of destruction, takes pride and pleasure in stocking our woodshed to provide solid warmth during the cold of winter. Our son, Gus, splits each piece with blows from a maul and wedge. My job is to cut kindling. Billy Collins expressed both the physical and emotional aspects of this process in his poem "Splitting Wood":

I want to say there is nothing
like the sudden opening of wood,
but it is like so many other things—

the stroke of the ax like lightning,
the bisection so perfect
the halves fall away from each other

as in a mirror,
and hit the soft ground
like twins shot through the heart.

And rarely, if the wood
accepts the blade without conditions,
the two pieces keep their balance

in spite of the blow,
remain stunned on the block
as if they cannot believe their division,

their sudden separateness.
Still upright, still together,
they wobble slightly

as two lovers, once secretly bound,
might stand revealed,
more naked than ever,

the darkness inside the tree they shared
now instantly exposed to the blunt
light of this clear November day,

all the inner twisting of the grain
that held them blindly
in their augmentation and contortion

now rushed into this brightness
as if by a shutter
that, once opened, can never be closed.

Although the firewood that we split and burn in our woodstove dissipates into ashes, another form of fuel that provides heat the world over is charcoal—burned wood that is burned again. It is created by charring wood very slowly, minimizing its contact with oxygen so that full combustion does not occur. Early on, the production of charcoal involved pit kilns, the amount of air being regulated by covering the burning materials with earth; later, when aboveground kilns became the norm, airflow was controlled by closing down vents. Charcoal makers, called colliers, tended as many as fifteen piles, each made up of twenty to fifty cords of hardwood.

Although today charcoal is mainly used for home barbecuing in the United States and as a domestic fuel for cooking and heating in developing countries, it persists as an important industrial fuel. Because it burns hotter and more efficiently than wood, it can generate enough heat to forge metals. Charcoal was first used in Europe as early as 5,500 years ago, and it fueled the Bronze and Iron Ages (2000–1000 B.C.). In some regions, half of the workforce was in some way involved in charcoal production to feed the forges of the ironmongers. About 25 percent of the world's fuel wood today—equivalent to 52 million tons per year—is converted into charcoal, large amounts of which are used in foundries and forges for the refining and shaping of metals and the production of glass; as a purifier of food and water; and in gunpowder.

WIND AND WINDBREAKS

> Who has seen the wind?
> Neither I nor you:
> But when the leaves hang trembling,
> The wind is passing through.
>
> Who has seen the wind?
> Neither you nor I:
> But when the trees bow down their heads,
> The wind is passing by.
>
> —*Christina Georgina Rossetti, "Who Has Seen the Wind?"*

The cloud forest of Monteverde, Costa Rica, is bordered by the San Luis Valley, a lovely rural community of forest and pastureland established in the 1960s. Over the next two decades, the number of pastures grew as demand for milk increased. Farmers noted that the productivity of the dairy cows decreased markedly in winter—between December and March—when the strong trade winds swept through the exposed pastures. In the early 1980s, farmers in San Luis began to plant trees provided by the dairy cooperative in Monteverde, most of which were exotic, fast-growing species such as cypress, eucalyptus, and casuarina. For the first few years, the tiny seedlings looked unable to do much of anything, but within five years they provided the dairy cows shelter from the wind. Meanwhile, the Monteverde Conservation League, a grassroots environmental group, focused on propagating native tree species for windbreaks, establishing two tree nurseries and growing saplings from seeds and cuttings, which they then distributed to farmers. By the late 1990s,

17. Windbreaks such as this colonnade of trees serve to protect plants and animals—particularly livestock—in tropical and temperate agricultural lands. Photo by Arp Kruithof.

over a million trees had been planted in Monteverde, and 70 percent of local farms were participating in the league's windbreak program.

The conservation group had to overcome considerable obstacles to get its program up and running. In 1990, windbreaks cost an estimated $800 per farm, an amount that covered the trees themselves, fencing to protect them from cows, herbicides to prepare the ground, fertilizers, and labor. Because it takes several years to realize any benefits from windbreaks, most farmers needed subsidies. With support from conservation agencies and local and international volunteers, the landscape now looks like a Through-the-Looking-Glass chessboard of pastures separated by healthy green lines of trees, protecting cows and people from the scouring winds.

A large literature exists on windbreaks, or shelterbelts as they are also called, defined as one or more rows of trees or shrubs planted as protection from the wind. They slow not only the speed of air but also whatever is contained in that air—moisture, smoke, particulates, pollution. Much of the literature documents how tree rows enhance biodiversity in agricultural systems. In England, for example, windbreaks take the form of hedgerows, narrow bands of woody trees and shrubs that thread between pastures, forming connective corridors that are critical for the movements of birds, small mammals, and invertebrates from one part of the landscape to another. Windbreaks significantly reduce the drying of pasture soils, at the same time improving meat and milk production, since cattle and other livestock need not expend as much energy to stay warm. In addition, windbreaks reduce soil erosion, provide habitat for wildlife and beneficial insects, and produce fruit and wood for farmers.

SHADE COFFEE

Shade is another sheltering contribution of trees that has drawn attention because of a single commodity: coffee. During a break at a recent meeting, a colleague and I bellied up to the espresso bar. Speaking in the nouveau jargon of the coffee-minded Pacific Northwest, I ordered a "double 2 percent grande latte with room, extra hot, one straw," then strolled over to the condiment shelf to spice it up with powered nutmeg and raw sugar. Although the venue and lingo—not to mention the cost!—would have been unimaginable fifty years ago, drinking a companionable cup of coffee is an activity that began over 1,200 years ago. Archeological evidence shows that coffee consumption spread along trade routes in Africa and Asia, becoming popular in Europe following its arrival in the seventeenth century. In the past two decades a thriving coffee culture has sprung up in the United States, and coffee can be obtained virtually anywhere—from tiny parking-lot kiosks to gas stations to laundromats. Today, Americans consume 300 million cups a day, a third of the world's coffee. One positive trend in this phenomenon is the growing demand for shade-grown coffee, thanks to an increasing awareness of the ecological impacts of coffee cultivation.

Coffee plants, which evolved in the shade of dense Ethiopian forests, do not grow well in the sun. In the past, coffee farmers raised their shrubs beneath banana, fruit, nut, and other hardwood trees. In the 1970s, however, scientists created hybrid high-yield, sun-tolerant coffee plant varieties that produce far more coffee per acre, but at a cost to the environment. Without the shade and nutrient replenishment of the forest canopy, these plants require larger amounts of pesticides, herbicides, and chemical fertilizers. Replacing indigenous forests with coffee plantations has long contributed to the alarming decline of suitable habitat for wildlife, including migratory birds that fly from North America to take up seasonal residence in tropical forests. Removing shade trees from those coffee plantations reduces even further the suitability of habitat for these birds.

In contrast, shade-grown coffee plantations mimic the natural environment of the wild plant. Practices range from "rustic" (using primarily old-growth and preexisting forest) to "planted" or "managed" (using shade trees planted specifically for the purpose of cultivating coffee). The most common tree planted for shade is *Inga edulis*, in the legume family. This species is able to symbiotically "fix," or add beneficial nitrogen to, the soil. Its branches also provide habitat for epiphytes and support birds with fruit and nectar. The deep-rooted tree improves soil and water quality and helps to prevent erosion on the steep hillsides where coffee grows.

Today, most of the world's shade-grown coffee farms are located in Central and South America. One example is Finca Irlanda, a 720-acre organic farm in Chiapas, Mexico. Coffee plants grow in the shade of seventy different tree species, including acacia, wild avocado, mango, balsa,

Mexican cedar, and *Inga.* These trees provide farmers with firewood, which reduces the need to harvest wood from surrounding forests for fuel. Together with more than a hundred species of understory and climbing plants, the trees also provide important habitat for over two hundred species of birds and almost eight hundred species of insects—including bees, their natural pollinators—which leads to healthier coffee shrubs and greater yields. Shade-grown coffee is often marketed under fair-trade conditions, making this product even more beneficial to humans. As awareness among coffee consumers increases, small-scale coffee farms will be able to thrive by using environmentally sound cultivation methods that not only sustain farmers and their communities but also preserve biodiversity in their home regions.

URBAN PARKS

City dwellers tend toward toughness, whether they are human youths, feral cats, or sidewalk trees. Urban trees are assaulted by air pollutants that damage their foliage, impairing photosynthesis and making them more susceptible to insect damage and diseases. Airborne pollutants include carbon dioxide (from burning oil, coal, and natural gas for energy), sulfur dioxide (from burning coal to generate electricity), hydrogen fluoride (from phosphate fertilizer production), ozone (from chemical reactions of sunlight on automobile exhaust gases), methane (from burning fossil fuels, livestock waste, and landfills), and chlorofluorocarbons (from air conditioners and refrigerators). As if that weren't enough, large areas within the city are capped by water-impermeable concrete and asphalt, leaving only tiny patches where water and air can pass to the cramped root systems of trees.

There is growing scholarly and political interest in urban parks—the postage stamps of nature that may contain less than a dozen trees and are surrounded by urban sights, smells, sounds, and pressures. Academic departments, journals (e.g., *Journal for Ecological Restoration*), and professional societies (e.g., International Society for Arboriculture) are now documenting the importance of city parks for humans. Until recently, planners and politicians largely viewed urban parks as public spaces useful mainly for recreation. Thankfully, they are now beginning to see them as much more. Chris Walker, a senior research associate with the Metropolitan Housing and Communities Policy Center in Washington, D.C., has articulated this broader assessment of parks, which encompasses urban policy objectives such as job opportunities, youth development, public health, and community building. Walker draws on statistically significant links between property values and proximity to green space in arguing for the importance of parks. In Philadelphia, for example, properties 2,500 feet from Pennyback Park realize an additional value of about $1,000 per acre, while those 40 feet from the park are $11,500 more valuable per acre. In Boulder, Colorado, the price of residential property—based on data from three neighborhoods—decreases by $4.20 for every foot removed from the greenbelt.

The total worth of urban parks, of course, goes beyond monetary value. Perhaps most important, they provide a venue where youth development can be fostered, as expressed in a report from the National Academy of Sciences. Urban teens, the report concludes, are best served by a constellation of community-based activities that place them at the center of neighborhood life, where they can engage with caring adults outside their families; learn rules of behavior, skills, and values; and cultivate a sense of personal identity.

One of the most innovative examples of this sort is the Empowering Youth Initiative, a program of Chicago's Garfield Park Conservatory, an urban botanical garden. Each year, a new group of fifteen fourth- through seventh-graders jointly designs a permanent display for the Elizabeth Morse Genius Children's Garden. The students work as real designers, exercising their brainpower and creativity in a team environment as they develop models, determine a budget, and collectively decide on the year's design. Thus, adolescents learn that they can associate positive intellectual challenges and social interactions with plants, parks, and people, and that their contribution to a particular piece of nature—their garden—

can have a powerful effect on the community as a whole. These insights enhance their sense of safety within society, indirectly fulfilling the second step of Maslow's modified pyramid, having to do with security.

Trees and parks also strengthen ties among community residents by bringing people together in informal ways, including those that are often divided by race or age. Research on low-income housing developments has found that parklike public spaces encourage residents to leave the isolation of their apartments and socialize; the lasting ties that they form contribute to the area's security and livability. The same is true on a larger scale. Prospect Park in Brooklyn, New York, for example, has a Community Advisory Committee, representing a wide range of interests, which meets monthly. The committee has strengthened the park's ties to the many ethnic and racial groups throughout Brooklyn, creating a strong sense of community and shared purpose.

PROTECTION, SECURITY, AND LOVE

> Do you still sing of the mountain bed we made of limbs and leaves:
> Do you still sigh there near the sky where the holly berry bleeds:
> You laughed as I covered you over with leaves, face, breast, hips and thighs.
> You smiled when I said the leaves were just the color of your eyes.
>
> Rosin smells and turpentine smells from eucalyptus and pine
> Bitter tastes of twigs we chewed where tangled woodvines twine
> Trees held us in on all four sides so thick we could not see
> I could not see any wrong in you, and you saw none in me . . .
>
> The smell of your hair I know is still there, if most of our leaves are blown,
> Our words still ring in the brush and the trees where singing seeds are sown
> Your shape and form is dim, but plain, there on our mountain bed
> I see my life was brightest where you laughed and laid your head . . .
>
> —*Woody Guthrie, "Remember the Mountain Bed"*

When I visit New York City, I stay with my dear friend Laurie Weisman in her co-op apartment in Greenwich Village. While there, my favorite outing is not to Times Square or Park Avenue or even the American Museum of Natural History, but rather to the narrow strip of a park along the waterfront on my daily runs. There, the broad paths are filled with sweaty joggers, moms with strollers, color-coordinated skaters, and elderly walkers. The trees that extend to either side of the paths give the park the semblance of a wildland island, essential for our connection to nature because of the ocean of concrete around us. After my run, I return to Laurie's apartment feeling uplifted, not just from the physical benefits of exercise, but from the knowledge that humans want and use even small stands of trees in their busy urban lives.

When humans feel that they are safe and protected, they are more inclined to reveal vulnerability. Trees provide glens and coves that serve as safe spots for lovers to exchange words of commitment for an instant or for a lifetime. The classic expression of enduring love is initials carved into the bark of a tree, which will exist for as long as the tree lives and grows. Even after the couple bearing those initials has died, tracings of their love will remain.

From the long-ago sleepovers in the tree house of my childhood and the busy-but-peaceful pathways of city parks I have visited in my travels since then, I have learned that trees provide elements of security that transcend the tangible elements of homes and shelter. Just recently, I visited Westwood Baptist Church in my hometown of Olympia with some friends who were visiting. Arriving a few minutes late, we slipped into the back pew to listen. The pastor was speaking on the need for all of us to seek and find an entity who will protect us, who will hold us in his arms to help us feel secure when we are frightened, and safe when there is danger around us. He spoke of being held by limbs that would support us forever, never tiring, never resting, helping us to find quiet and calm in our lives. I was amazed and pleased that the pastor would include a description of trees and their spiritual benefit to humans in his sermon. I recalled a line from a William Stafford poem: "I rock high in the oak—secure, big branches—at home while darkness comes." It was only at the end of his sermon that I realized he was talking not about trees at all, but about Jesus. I realized that he and his flock view Jesus the way I view trees, as entities that hold us in their strong limbs, providing a safe refuge whenever we may need it.

The Presence of Trees

I have always felt the living presence
Of trees
The forest that calls to me as deeply
As I breathe
As though the woods were marrow of my bone
As though
I myself were tree, a breathing, reaching
Arc of the larger canopy
Beside a brook bubbling to foam
Like the one
Deep in these woods,
That calls
That whispers home

—*Michael S. Glaser*

Chapter Four

HEALTH AND HEALING

> When we are stricken and cannot bear our lives any longer, then a tree has something to say to us: Be still! Be still! Look at me! Life is not easy, life is not difficult.
>
> —*Hermann Hesse, "On Trees"*

I heard drums and the high, sweet "yi-yi-yi" of women singing outside my small hut in the highlands of Papua New Guinea. It was a balmy tropical night, and the villagers were celebrating the birth of a child with food, music, and dancing. I was huddled under my covers, shivering violently. It was 1976. I was twenty-two years old, ten thousand miles from home, working my first job in a remote field station in the tropics collecting plants for a biologist studying plant-insect interactions. I had malaria. Sometime in the weeks after my arrival in Papua New Guinea, a female *Anopheles* mosquito bit me. At the site of her bite, she left a few protozoa of the genus *Plasmodium* inside my skin. She had picked up these hitchhikers from another person infected with malaria, and the microscopic invaders spent a week burrowed in her gut before moving to her salivary glands for transmission to their next human host—in this case, a recent college graduate from Maryland eager to explore the tropics. The parasites then lodged in my liver, and after a few days of development, invaded my red blood cells, reproduced, and exploded them, pouring their progeny into my bloodstream to continue the cycle.

The word *malaria* comes from medieval Italian: *mala aria*, or "bad air."

The English called it "marsh fever." It has been a scourge of humans for centuries. Twentieth-century efforts to eradicate the disease successfully suppressed it in many temperate regions—sometimes with unintended negative effects, such as those that resulted when DDT was introduced to the environment. However, up to three million people still die from the disease each year, mainly in the tropics, young children and pregnant women being the most vulnerable. But I did not die. A week after my initial bouts of shivering and fever I had recovered and was back at my tasks as a field assistant. I owed my recovery to quinine, a drug that human beings have used to combat malaria for centuries—a remedy that comes from trees.

Quinine is derived from the bark of *Cinchona officinalis*, a small-statured tree with large glossy leaves and arrays of fragrant pink flowers. Native to the South American Andes, *Cinchona* is in the family Rubiaceae, along with coffee and the sweet-scented gardenia. The name quinine comes from the Quechua word for the tree's bark, *quina-quina* (literally, "bark of barks"), which the local inhabitants dried, ground into a fine powder, and mixed with water for relief of malaria. The plant reportedly takes its scientific name from the countess of Cinchon, wife of the viceroy of Peru, who in 1638 was cured of malaria with this decoction.

In its pure state, quinine is a white crystalline alkaloid. It combats malaria by interfering with the parasite's ability to break down and digest hemoglobin in the blood. In Western medicine, large-scale use of quinine to prevent malaria started around 1850, spurring widespread harvesting of the bark and decimation of the *Cinchona* in its native forests. Because the trees were so valuable, South American officials prohibited their export. In the 1860s, however, British and Dutch entrepreneurs smuggled seedlings out of Bolivia and established extensive *Cinchona* plantations in Java and Sri Lanka. By the 1920s, these two countries dominated the world production of quinine. The drug generated huge profits for its European growers and importers, though little of that wealth returned to people in South America. In the 1940s, chemists successfully produced quinine in the laboratory, and since then, synthetic alternatives such as chloroquine have reduced reliance on natural quinine. However, there has been a troubling rise in chloroquine-resistant malaria since the time I shook on my cot thirty years ago in Papua New Guinea, and today, quinine from *Cinchona* is still used to treat certain highly resistant strains of malaria.

Quinine has also been used for flavoring, especially to provide tonic water with its signature bitterness. In fact, gin and tonic—the cocktail once ubiquitous in India and other tropical British colonies—was originally imbibed to prevent malaria attacks, though the tiny amount of quinine contained in tonic water is actually too small to have an effect. Quinine bark is harvested today much as it has been for hundreds of years. Tree trunks are beaten and the peeling bark is removed. Although the bark can partially regenerate on the tree, after several cycles of bark removal new trees are planted. About 300 to 500 tons of quinine alkaloids are extracted annually, about half for beverages and the remainder for prescription drugs. And malaria is only one of many diseases for which trees hold the remedy.

TREES AND MEDICINES

For centuries, trees have participated in the treatment of ailments ranging from diabetes to depression. The chewed leaves of the South American coca bush, for example—despite the plant's dark reputation for its most notorious product, cocaine—are an important aid in dealing with the physical stress of living at high elevations. This densely leafed plant native to the eastern slopes of the Andes grows as a large bush or small tree and can be harvested for twenty years or more. In Western medicine, the alkaloid derived from coca leaves has given us the chemical blueprint for human-made substances used in local anesthesia, including procaine, lignocaine, and novocaine. And a West African vine, *Strophanthus kombé*, which climbs up the trunks of trees and hangs in coils between them, contains strophanthin, a glucoside that effectively combats heart disease, with fewer gastrointestinal side effects than occur with other drugs prescribed for coronary problems.

Not only shrubs and vines are used as medicines. A tall deciduous African tree that grows in dry forests, *Prunus africana*, has for centuries been used by traditional healers for a variety of illnesses, including disorders of the prostate gland. With the growing number of aging men in Western populations and our growing confidence in natural medicines, the pharmaceutical industry has capitalized on this source of tree-derived healing. The harvested bark currently brings in $200 million each year, and that value is rising, along with concerns for its conservation (as we will discuss later).

Practitioners of traditional, or Ayurvedic, medicine in India have long turned to trees for source materials. Indeed, when my father came to the United States in 1947 as a doctoral student, it was to investigate the potential efficacy of India's native mayapple plant (*Podophyllum peltatum*), an ephemeral herb that grows in the shade of deciduous forests. Arboreal members of the Ayurvedic tradition include the bark of the wild cinnamon tree, or karuva (*Cinnamomum zeylanicum*), which relieves digestive problems such as cramping and gastroenteritis. Baheda (*Termanlia belerica*), a large deciduous tree with an umbrella-like crown, is found all over the Indian subcontinent; its fruits are used to relieve sleep disorders and high blood pressure, and to treat leprosy. A tree species of the Western Ghats (Sahyadri Mountains) in the southwestern corner of the country, *Nothopodytes nimmoniana*, is gaining international attention because one of its extracts, camptothecin, has promising anticancer benefits.

Almost half of all prescription drugs dispensed in the United States contain substances of natural origin. Who discovers these healing properties? Given the huge plant diversity in forests, how are particular tree species identified as holding the next cure for diabetes, the common cold, or baldness? And how do plants promote our mental health? If we were to host a dinner with specialists on trees and healing, we would need a very large table. There would be a plant taxonomist, an Amazonian shaman, an Algonquin elder, an organic chemist, a city arborist, an urban planner, an environmentalist, and a counselor for people recovering from emotional trauma and drug addiction. Each has unique insights into the ways in which trees enhance human health. The head of the table would be reserved for that rare breed of researcher, the ethnobotanist, an interdisciplinarian who combines plant knowledge, pharmaceutical expertise, cross-cultural communication, and entrepreneurship to find, extract, and disseminate medicinal compounds from trees.

MEDICINAL TREES OF THE TROPICS

The sheer diversity of tropical forests, which can support more than 180 tree species per acre— double that when we include herbs, mushrooms, ferns, shrubs, and epiphytes—makes the job of finding useful plants a formidable one. In 1988, a time when biologists were just becoming aware of tropical deforestation, I attended a conference on tropical rainforest conservation at the California Academy of Sciences. One of the participants, Mark Plotkin, talked about his ethnobotanical research. He concluded that we do not always need to seek new knowledge; rather, we should learn what the people who live in rainforests, particularly the medicine men and shamans, already know. Of the hundreds of pharmaceutical drugs that are plant derived, he said, three-quarters were discovered thanks to information collected from indigenous peoples. "Each time a medicine man dies, it is as if a library has burned down," Mark said. "We often talk about disappearing species, but the knowledge of how to use these species—especially healing plants—is disappearing much faster than the species themselves."

Today, most shamans are elderly. Younger generations are losing interest in this knowledge, lured away by the modern world. For example, among the Tirio people, a tribe on the Brazil-Surinam border that uses about three hundred plants for medicine, the few shamans who remain all exceed sixty years in age, and none has an apprentice. In the early 1990s, Mark responded to this situation by initiating what he calls a "sorcerer's apprentice" program, which encourages young tribal members to learn the medical knowledge of elderly shamans. He plans similar efforts for tribal communities in Costa Rica, Colombia, Ecuador, Guyana, and Surinam.

18. Shamans of the Amazon and other tropical and temperate forests hold vast and important knowledge of medicinal and other useful plants. Reproduced by permission of the Amazon Conservation Team.

Ethnobotany has given rise to the practice of "bioprospecting," in which botanists work with rainforest shamans to learn their plant knowledge and then connect plants with manufacturers and marketers of drugs. It is a high-stakes endeavor. In the United States, drugs whose active ingredients are derived from tropical plants are worth $25 billion each year, according to the Millennium Ecosystem Assessment project of 2005. Ethnobotanists working with economists estimate that yet-to-be-discovered drugs are worth $109 billion to society as a whole. Like the California gold rush, this "gene rush" has the potential to benefit or destroy fragile ecosystems, as well as the people living in them. At its best, bioprospecting allows representatives of both cultures to target plants with medical applications, synthesize their constituent chemicals in a laboratory, and patent them for drug manufacture. At its worst, it steals indigenous knowledge only to benefit stockholders and bolster academic careers.

I witnessed one act of bioprospecting up close in 1995 during a summer of fieldwork in Monteverde, Costa Rica. My next-door neighbor there, Will Setzer, a chemist, was collaborating with an ecologist, Bob Lawton, to identify potentially useful natural products from local plants. Their previous research informed them that up to 10 percent of the dry weight of plants was made up of secondary chemicals that evolved to defend against infection or insect damage. There was a good chance that some of that defensive material could be useful as medicines. Because Monteverde lacked any sort of laboratory, Will set up an apparatus to extract plant chemicals on his own front porch. My family and I would often stroll over in the evening to watch his potions bubble and drip to purity in the homemade distillation device—one not so different from a moonshiner's still in the backwoods of Prohibition-era America.

To distill a plant is a complex and time-consuming task, so only a handful could be tested. With over three thousand species of vascular plants in the cloud forest to choose from, how did these two scientists decide which ones to test? Because aboriginal people do not live in the cloud forests of this area, Monteverde has little ethnobotanical tradition to draw from, so Mark Plotkin's approach of interviewing a local shaman was impossible. Instead, Will and Bob began with plant families that have pharmaceutically active members in other regions. Initially, they concentrated on the ginseng family, some members of which have been used in traditional Eastern medicine. In tropical cloud forests, this family is represented by trees as well as herbs. These two modern-day shamans made crude extracts from leaves of twelve species to screen for antimicrobial activity and for in vitro anticancer activity (toxic activity against cancer cell lines, in culture). Six of the twelve showed "remarkable cytotoxic activity" against cancer cells without harming normal blood cells. When they returned to their laboratory in Alabama, Will and his students were able to decipher the structure of the active compounds and publish papers on the chemical constituents of possible drugs.

Was a cure for cancer being discovered on my neighbor's porch even as we discussed the arrival of the three-wattled bellbirds from the lowlands and dessert options for dinner later that night? Will shrugged and said that the bottleneck for getting drugs from forest to market lies not in the identification of active plants, but rather in the development, testing, and marketing of new drugs. He flipped through the pages of a journal full of scientific articles with the names and chemical structures of healing compounds. "That," he said, "is where most of this work stops." Rigorous testing requirements of regulatory agencies such as the Food and Drug Administration have resulted in an enormous backlog of drugs that might be able to cure illnesses waiting for approval. "Orphan diseases"—those that afflict either very few people or very poor people—have a far smaller chance of gaining the benefits of plant-derived remedies because of the expense of this process. Nonetheless, Will and his colleagues keep trying and testing, in the hope that one of these compounds, one day, will reduce human suffering.

Tropical medicinal plants play particularly important roles for rural people in developing countries, 90 percent of whom rely on traditional medicine for their primary health care. Researchers from the New York Botanical Garden, the center of North American ethnobotany, have reported that if the medicinal plants, fruits, nuts, and oils are harvested sustainably, rainforest land has much greater immediate and long-term economic value than if it is clear-cut for timber or converted to pasture. Whereas conversion of land to cattle operations yields the landowner $60 per acre, and $400 an acre if timber is harvested, land used to produce such sustainable resources as medicines, foods, and rubber can yield the landowner $2,400 per acre, providing a strong incentive for people to protect their forests.

MEDICINAL TREES OF TEMPERATE ZONES

"It will beggar a doctor to live where orchards thrive," says an old Spanish proverb—the Iberic version of "An apple a day keeps the doctor away." As in the tropics, many trees native to temperate regions of the world provide materials for the treatment of diseases. Most were developed by indigenous people, and ethnobotanical studies have generated vast lists of medicinal plants—by species, by ailment, by location. Here are but a few, from the forests of the northern continental United States.

Red alder, one of the most common trees of the Pacific Northwest, thrives in wetlands and on riverbanks. It is celebrated by ecologists and foresters because it adds nitrogen, a required nutrient often in short supply, to the soil. Alder trees have a mutualistic relationship with microbes that live in small grapelike clusters attached to the tree roots. These microbes take atmospheric nitrogen—a form unavailable to plants—and convert it into ammonium, a form that plants can readily absorb. The microbes get a safe place to live, and the roots of the alder get a fertilized front yard.

This tree and other species of deciduous trees have healing properties. Red alder bark can be simmered in water to make a healing wash for wounds that pulls the edges of a cut together. The spring growth of the twig tips and compound leaves of ash can be boiled to make a laxative tea that eases symptoms of rheumatism. The smooth gray bark of the beech tree is used as a tea for lung problems and as a soothing wash for poison ivy. In the Great Lakes region, the Ojibwa made a decoction of the inner bark as a wash for sore eyes. The tannin in the bark and leaves of white oaks is an antiseptic.

Many evergreen species have healing properties also. Northern white cedar, which grows throughout the northern Midwest, is used by the Algonquin, who spread its branches on the floor of their sweat lodges and distill a tea—high in vitamin C—from its scalelike leaves. Chinese herbalists use a fungus, *Poria*, that grows on the roots and trunks of white pine as a sedative. The tamarack tree exhibits a mixture of coniferous and deciduous habits, producing cones but dropping its needles in winter and growing new ones in the spring. The Ojibwa—who call tamarack *ëmuckigwatigí*, meaning "swamp tree," as it often grows in bogs—treat headaches by crushing its leaves and bark, placing them on hot stones, and inhaling the fumes.

The most famous medicinal conifer is the Pacific yew, a tree I encounter when I go trail running along a beautiful mountain path known as the Staircase, on the eastern slopes of the Olympic Mountains. My favorite spot on this route hosts a large rock and a single yew tree. It is easy to overlook yews in the forest; they are small, nondescript trees that grow in the dark understory. Their modest size can belie their age, however, for they grow in tiny increments for centuries. Neither stately nor particularly symmetrical, the unassuming yew has nevertheless provided one of the most successful cancer drugs ever developed, Taxol, which comes directly from its bark and has been found to be amazingly effective against ovarian cancer.

How did this modest tree become the poster child for plant-derived medicines? In 1958, the National Cancer Institute initiated a program to screen 35,000 plant species for anticancer activity, enlisting the U.S. Forest Service and a brigade of academic botanists to collect samples. Collaborating pharmacologists soon found that yew-bark extracts interfered with the growth of cancer cells. In 1971, the active ingredient in the bark, paclitaxel, was identified as one of a family of drugs called antineoplastics. Clinical trials of this compound against numerous cancers showed a 30 percent response rate in patients with advanced ovarian cancer. Similar to other antitumor pharmaceuticals, this drug has side effects, including a low white blood cell count, nausea, loss of hair, and lower resistance to infections. However, it received clearance from the Food and Drug Administration in 1992, and pharmaceutical companies now synthesize the drug in their laboratories. By 1996, survival time for women with advanced ovarian cancer had been extended by 50 percent over women using standard therapies. Each time I run up the Staircase trail, I pause at the little yew tree by the rock and offer my thanks for its healing capacity.

THE HEALING ENVIRONMENT

A beloved O. Henry story of my youth was "The Last Leaf." The plot involves a young woman named Johnsy who, having contracted a near-fatal case of influenza, decides that she will die when the last leaf of the ivy vine outside her sickroom window falls to the ground. An impoverished artist named Mr. Behrman sees to it that this doesn't happen, painting on the wall a leaf "that never fluttered or moved when the wind blew"—and then dying himself of pneumonia after exposure to the rain and cold. His ruse works, however; as the artist draws his last breath, Johnsy declares her intention "some day . . . to paint the Bay of Naples," having regained the will to live.

The curative value of greenery outside a sick person's window is not just the stuff of fiction. Roger Ulrich, a professor of behavioral psychology at Texas A&M University and a pioneer in the study of environmental influences on health, contends that having nature close by can enhance human well-being in measurable ways. In his view, no environment is neutral, and the surroundings in which a hospital patient receives care significantly affect outcomes in both positive and negative ways. Fear and uncertainty about the prognosis, isolation from friends and loved ones, and related stresses can lead to suppression of the immune system, as well as dampening emotional and spiritual energies, and thus impede recovery. In the early 1990s, Ulrich observed that the psychological and social needs of patients had been largely disregarded in the design of health-care facilities. Rather than providing an environment that calms patients and strengthens coping resources and healing processes, our "state-of-the-art" hospitals are frequently stark and impersonal, stressful to patients and detrimental to caregivers.

Ulrich went beyond theory and tested these ideas. In a landmark study published in the journal *Science*, he investigated the recovery rates of a group of patients with contrasting views from their windows in a Texas hospital, all of whom had had gall bladder surgery. Some patients had rooms that overlooked a patch of trees in the hospital courtyard; others had windows that faced a concrete wall. Patients in rooms with views of trees spent fewer days in the hospital, used fewer narcotic drugs, had fewer complications, and registered fewer complaints with the nurses. Other studies showed that environments with nature-related imagery, such as photographs and paintings on the wall, reduce anxiety, lower blood pressure, and reduce pain. Not only did Ulrich demonstrate a direct correlation between having a view of trees and better health, but research has since shown the obverse to be true as well, linking unsupportive surroundings such as blank walls and harsh lighting to elevated depression, greater need for pain drugs, and longer hospital stays. Even the most mainstream administrators, hospital architects, and interior designers now acknowledge that environments enhanced by plants and nature imagery—from reception to radiology to restrooms—empower patients in their healing, provide relief for worried families, and fortify care providers.

Of course, not every hospital or clinic is fortunate enough to have views of a forest. Over the past decade, therefore, a number of private companies have been marketing artificial "treeviews" to healthcare facilities. One such company, SkyFactory, offers two product lines. The SkyCeiling, which creates a photographic illusion of a real tree canopy against a cloud-dotted blue sky, is installed in ceiling grids where fluorescent lights and acoustic panels usually reside, while the Luminous Virtual Window features landscape images for vertical walls. Though relatively expensive—panels cost between $85 and $100 per square foot—they are easily installed, with light sources radiating 6500K "daylight" into the room. These companies post enthusiastic testimonials from patients about how views of blossoming and leafy trees had a soothing and balancing effect on them. Although little quantitative information exists to attest to their direct health benefits, the fact that the artificial treeview business has grown exponentially in the past five years suggests their effectiveness—enough so, anyway, that hospital administrators are willing to pay handsomely for a room with a view, albeit a manufactured one, of trees.

Humans react to particular tree shapes. Sociological research by Virginia Lohr and Caroline Pearson-Mims at Washington State University explored the effects of specific tree shapes and forest structures on people. They based their work on research from the 1980s which demonstrated that people have powerful, positive aesthetic, emotional, and physiological responses to scenes of natural landscapes. For example, in a study conducted in a typical windowless work environment, participants were more productive and less stressed when live interior plants were added to the room. Another study showed that university students who watched a stress-inducing movie followed by videotapes of nature recovered from the stress more quickly than people who watched videotapes of urban scenes following the movie.

Related research revealed that humans respond more positively to trees than to other plant types in the landscape. People also respond differently to tree canopies of different colors, finding those with the green hues associated with healthy trees more appealing. Lohr and Pearson-Mims also tapped the "savanna hypothesis" of University of Washington zoologist Gordon Orians. He theorized that, because of our evolutionary origins in the savannas of Africa, humans should highly prefer a savanna-type landscape, with its widely spaced trees with low, spreading crowns. That environment, after all, provided the essentials for human survival: food, trees for protection from predators and the elements, and long, unobstructed views for observing both predators and prey. Even vegetation designed for purely aesthetic pleasure, such as that in parks and gardens, tends to reflect these evolutionarily based preferences, Orians noted; through genetic selection and pruning, humans go to great pains to make trees resemble savanna forms rather than locally wild forms. In Japan, for example, the ideal tree for a formal garden is broad relative to its height, with a short trunk and small, divided leaves. Trees that do not naturally exhibit this growth form, which is typical not of Japanese pine forests but of the African savanna, either are not used in formal Japanese gardens or are heavily manipulated to assume a savanna form. To test the savanna hypothesis, Lohr and Pearson-Mims showed over two hundred participants scenes with inanimate objects as well as different tree forms—spreading, rounded, or columnar—in two urban scenes. As predicted, participants found scenes with trees more attractive than scenes with inanimate objects, and they rated spreading trees more appealing than rounded or columnar trees. Thus, human sensitivity to even very subtle differences in forest structure seems to have roots deep in our own evolutionary history.

HOLISTIC HEALING

In Kos, Greece, in front of the Castle of the Knights, stands the Plane Tree of Hippocrates, a huge tree whose crown spans a perimeter of nearly forty feet. Legend has it that Hippocrates himself planted it and taught in its shade. On a nearby plaque, visitors can read the Oath of Hippocrates that physicians take when they enter the medical profession. One section reminds doctors that healing requires more than chemistry and mechanics. A modern version by Dr. Louis Lasagna states:

> I will remember that there is art to medicine as well as science, and that warmth, sympathy, and understanding may outweigh the surgeon's knife or the chemist's drug. . . . I will remember that I do not treat a fever chart, a cancerous growth, but a sick human being, whose illness may affect the person's family and economic stability. My responsibility includes these related problems, if I am to care adequately for the sick.

Although this personal side of medicine has declined over the past century, recently more holistic approaches to caregiving have been reentering the health profession, reaffirming that oath. Some universities, such as the University of Virginia, now offer "health and spirituality" classes to their medical students. Counselors and humanists are being added to medical teams, and chapels and chaplains have a more visible presence in hospitals.

Part of the growth in holistic medicine involves visualization therapy, or guided imagery, a form of self-hypnosis that promotes physical, emotional, and mental health through the imagining of desired outcomes to specific situations. Imagery has been a healing tool in many cultures. Navajo Indians, for example, practice a form of visualization in which the person "sees" himself as healthy. Although the American Medical Association has not sanctioned these techniques, millions of Americans practice some form of imagery or meditation to reduce stress, boost the immune system, or cope with life-threatening illness, and they do so not sitting pretzel-like with a guru on a mountaintop in the Far East, but right at home, in hospitals, schools, law firms, corporate offices, and prisons. Many of the nation's major cancer centers now offer visualization therapy. How does it work? Cancer patients are taught to visualize their white blood cells as sharks, for example—numerous and powerful, attacking and destroying the mutant cells, which might be imagined as ugly fish. On a neurological level, guided imagery changes brain chemistry, increasing the flow of alpha brain waves and stimulating the release of neurochemicals such as dopamine. It certainly appears to improve patients' attitudes. In a five-year study published in the *New England Journal of Medicine* in 2001, researchers at the University of Toronto found that breast cancer patients who attended weekly visualization therapy groups, in addition to undergoing conventional chemotherapy, reported much less anxiety and pain than patients who went through standard treatment only.

How does all of this relate to forests and trees? Although most visual imagery therapies use animals, such as tigers and sharks, to represent action, strength, and destruction, trees can serve as evocative, empowering symbols as well. I was struck by this recently while visiting a tiny clifftop cabin our family owns on the rugged west coast of Washington State. The land borders on Olympic National Park, and a two-minute walk takes us down the steep cliffs to the Pacific's pounding waves. The cabin barely nestles within a stand of coastal Sitka spruce, whose growth form expresses the harshness of their habitat, their tiny crowns having been sculpted into a permanent "brush cut" by the nearly constant salt-laden winds. Most of the trees sport enormous burls as well, which grow slowly and irregularly—but inexorably. A burl begins life as a gall, a tumor of plant tissue stimulated into existence by fungi, insects, or bacteria; ultimately the gall becomes a woody globose structure, sometimes as large as a bathtub, which affects the tree's wood structure but does not kill it.

As I rambled into the forest, I noted several snapped limbs lying on the ground, severed during a windstorm the previous night. Although viscous sap oozed from the wounds on the trunk where the branches had broken, I knew the tree would survive its loss, since the trunks of nearly all the trees around me were punctuated by knots, evidence of prior limb

19. The burls—or undifferentiated growth—of this Sitka spruce tree can inspire guided image therapy for cancer patients. Photo by Scott Hollis.

loss. In fact, many trees adapt to the loss of photosynthetic apparatus that comes with the breaking of a limb by sprouting so-called epicormic branches, which arise when sunlight stimulates the cambium—the inner ring of growing wood—to produce a new stem that takes on the form and function of the original branch.

Walking slowly among these trees that blustery morning, I recognized that despite the heavy winds and evidence of tree damage all around me, in the form of wind-shorn crowns, giant burls, and snapped limbs, this stretch of spruce forest provided me with a profound sense of calm. I thought about how the trees' remarkable persistence—manifested in crowns bent and scoured, yet carrying on with the daily business of photosynthesis, cone making, and root elongation—could be useful for people who face analogous challenges in their external and internal landscapes. Specifically, I wondered whether people who struggle with chronic illnesses might gain from images of forests like this wind-battered spruce woodland to reduce stress and evoke a sense of hope.

When I returned home and explored the literature on visualization therapy, I found that several health groups use tree and forest imagery to help patients with debilitating pain and life-threatening illnesses, mainly to complement other, more conventional therapies. For example, Dan Johnston, assistant professor of behavioral science at Mercer University School of Medicine in Georgia, noted that the stress of chemotherapy or radiation is often accompanied by physiological changes such as high blood pressure and a rapid heart rate, as well as emotional reactions such as feeling tense, apprehensive, or frustrated. These can in turn aggravate some of the potential side effects of treatment, such as nausea, fatigue, and low energy. A patient who approaches her treatment procedure in a relaxed state of body and mind will reduce the likelihood of such side effects occurring. One way of achieving this state, said Dr. Johnston, is through use of nature images.

Another group, the Empowered Within Project, draws on the quiet power of trees to help children cope with serious illnesses. CDs are used to prepare the child, both physically and emotionally, for treatments of chemotherapy. In addition to "Your Friendly Dragon" and "Astronaut Blast-off," there is a track called "The Strong Tree":

> This is a story where you will use an image of a tree to give you a picture of strength that you can use whenever you wish to be strong. Sometimes a serious illness or a medication can have side effects that cause our bodies to change. This story helps remind us that even with changes in our appearance, we are still the people we were before, just like the tree is still a strong tree even when one of its branches fell down.

TREES AND EMOTIONAL HEALTH

In addition to providing us with physiological and psychological support, trees give important spiritual and emotional sustenance. In her book *Myths of the Sacred Tree*, Moyra Caldecott writes:

> For our health, we have to preserve our physical world—the environment our bodies need to survive on this earth. But we also must preserve the environment our spirits need to survive in eternity. The tree is essential to one, and the symbol of the tree is essential to the other. The physical tree sustains our

> bodies with its fruit, its shade, its capacity to reproduce oxygen and to hold the fertile soil safe. The mythic tree sustains our spirit with its constant reminder that we need both the earth and the sunlight—the physical and the spiritual—for full and potent life.

For centuries, humans have looked to the symbolic resonance of nature to mark the cycles of life and to heal our sense of loss from death, the ravages of war, and natural disasters. One of the most direct applications of the association of trees with human emotions is the Healing Trees Project, part of the remarkable Living Memorials Project that grew up in response to the events of September 11, 2001. To honor and memorialize people and communities who sustained losses, Congress asked the U.S. Forest Service to create an initiative that would invoke the healing power of trees to bring people together and soothe their sorrow. The project soon spread to agencies at the state and local levels, and a National Registry was created that now keeps track of hundreds of community-based, living memorial sites.

Let's look at three tree species—the dawn redwood, American elm, and weeping willow—to see how the project creates meaning. The first, the dawn redwood, is a deciduous conifer, with soft needlelike leaves that are shed in winter. This tree was discovered first through fossils unearthed in the early 1940s in Japan. Though thought to have been extinct, living specimens were soon found in China, and the tree is now a common ornamental. The dawn redwood symbolizes the idea that it is always possible to renew our place in the world and resume vital roles in our communities.

The second species, the American elm, was once a common and graceful urban street tree. Then in the 1930s, Dutch elm disease, caused by the fungus *Ophiostoma ulmi*, destroyed two-thirds of the population. A program of care has partially returned the elm to health through cross-breeding with highly resistant progeny. This recovery is a chronicle of how healing can occur through active care across generations.

The third species, the weeping willow, has graceful downward-arching branches, associating it with gentler, more inward aspects of human emotions. Willow leaves have been used by Native Americans to reduce fevers and headaches, and the active ingredient, salicylic acid, led to the creation of aspirin. The downcast branches of willows remind us that grieving is a natural part of the healing processes, as William Shakespeare reminded us as well:

> The fresh streams ran by her, and murmur'd her moans,
> Sing willow, willow, willow; Her salt tears fell from her,
> and soften'd the stones, Sing, willow, willow, willow.
>
> —*Othello* 4.3.43–46

Another group uses tree imagery to heal damage of a different sort—domestic violence, one of the most devastating and least witnessed types of harm in the world. Since 1991, A Window Between Worlds (AWBW), a nonprofit organization based in Venice, California, has offered refuge to thousands of women and their children in crisis shelters and transitional homes. But they go a step further, offering art workshops to promote healing and a renewed sense of hope. AWBW believes that through art, survivors can learn to see past the abusive messages they heard from their batterers, rebuild their self-confidence, and transform how they view themselves. One of the AWBW projects features trees as healers: "A tree is a powerful symbol of life. It gathers strength from the help it receives from its environment and it never stops growing or changing. It adapts to seasons and it carries with it an imprint of its life experiences. Through the Story Trees Project, our organization welcomes all survivors of domestic violence to join us in sharing our journeys."

Not every type of emotional damage is on the scale of terrorist attacks or domestic violence. Our day-to-day lives have their share of emotional dents and dings, and even here, trees can be mustered to repair or at least alleviate the hurts. Several years ago a Japanese forest ecologist, Shoji Takimoto, visited our Canopy Lab at the Evergreen State College. His visit coincided with our weekly lunch, and he joined my students and me as we described the progress of our projects. When we had finished, Dr. Takimoto told us in his quiet voice about a Japanese custom called *shin-rin yoku*, which translates literally to "tree shower." When a person feels anxious or tense, he simply goes outside, walks up to a tree, raises his arms above his head, and receives that which the tree provides—a sense of calm, serenity, and well-being. Ever since Dr. Takimoto's visit, the phrase "you should go take a shin-rin yoku" has been frequently heard in the Canopy Lab.

Like Dr. Takimoto, many humans associate nature with deep emotional healing, to mend past trauma, perhaps, or to recover from an addiction. When I browse the self-help section of a bookshop, I am amazed at the number and variety of approaches there are to self-betterment. Some advocate following the patterns of individuals who have been extraordinarily successful in overcoming hardship; others embrace the principles of established spiritual practices. Gail Faith Edwards, in *Opening Our Wild Hearts to the Healing Herbs*, writes: "Throughout time, humans have looked to trees for help in understanding their personal growth and changes. People have always sought solace and comfort from trees, spoken to trees of their problems and joys, and gathered strength from these interactions with the tree spirits. A common practice in many cultures is choosing a tree as a personal friend and ally. Those who do this learn lessons from and grow with that tree throughout life."

Trees offer inspiring examples of recovery and resilience. Many can germinate and survive in infertile terrain. Their seeds can remain dormant until conditions become suitable for growth. They exemplify steadfastness and courage—important qualities in any process of emotional healing. Trees, as Lynn Martin expresses so beautifully in her poem "Under the Walnut Tree," give us comfort even if we cannot articulate the source of this relief:

When I face what has left my life,
I bow. I walk outside into the cold,
rain nesting in my hair.
All the houses near me
have their lights on. Somewhere,
there is a deep listening.

I stand in the dark for a long time
under the walnut tree, unable
  to tell anyone, not even the night,
what I know. I feel the darkness
  rush towards me, and I open my arms.

I have reflected on how trees can model a pathway to health for those suffering from emotional loss, trauma, or addiction. First, trees teach us that it is possible to modify our environment, or reverse the effects of past environments. The soil in which a tree grows has much to do with its health. Soil is created from parent material, the raw geological matter laid down in horizontal strata across eons. Over time, weathering and aeration can modify and mix these layers, improving the tree's access to nutrients and water. The tree itself can bring changes by drawing nutrients from deeper in the soil, incorporating them into its foliage, and then depositing its leaves on the forest floor, where they decompose, becoming available again as food. In a similar way, humans must pay attention to their own substrate: their genetic makeup, the outlooks and actions of others that support or do not support their growth, and the attitudes that they have inherited. For many people with emotional problems, the early substrate, or baseline material, was characterized by fearful, deceptive, or violent interactions. Yet just as some trees can alter the fertility of their soil, humans can amend the medium in which they live by surrounding themselves with the nourishing fertilizer of supportive friends and community.

Second, all living things need energy. Trees gather energy from the sun through the process of photosynthesis. Numerous adaptations—such as the arrangement of leaves on their axils and the leaves' ability to move to maximize exposure to the sun at different angles during the day and during the year—allow a tree to harvest every available photon. Humans derive their energy not only from food, but also from feeling inspired and creative. Sometimes people lose contact with sources of joy that keep them engaged with life. Just as trees have mechanisms to maximize sunlight capture, humans must bring inspiration to their lives, by purposefully seeking out contact with others or creative projects.

Third, trees can teach us how to tame the chaos of our own lives. Transformation, even if it is from emotional turmoil to calm, can create a feeling of discomfort and disorder. The predictable and orderly growth pattern of trees—which we can see everywhere—can reassure us that order in our own lives is possible as well. Fourth, trees, rooted as they are in the earth and reaching ever skyward, exemplify a sort of grounded connectivity. Often, people under stress find themselves pulled from task to task and conversation to conversation, never finishing one before another demands attention. It is rare for them to be still; rarer still for them simply to be. Trees can provide inspiration to be silent and still. The "tree pose" in yoga exemplifies this, as does the Zen practice of *zazen* (literally, "seated meditation"). From trees, we learn to stop and sit, breathe in, breathe out, and see that movement is not always progress, and action not always accomplishment. With that silence can come the space needed for reflection on who we are, as John Ashbery describes in his poem "Some Trees":

> . . . you and I
> Are suddenly what the trees try
>
> To tell us what we are:
> That their merely being there
> Means something; that soon
> We may touch, love, explain.

Finally, trees embody the intriguing paradox of being strong and fragile at once. On the one hand, they exemplify longevity and hardiness, reminding us that the root of the word *tree*, *dāru*, is found also in the word *endurance*. They endure storms, wind, disease, and fragmentation of their environment. On the other hand, they can be extremely sensitive to disturbance. Increases in atmospheric concentrations of lead as small as one part per million, for example, can be fatal to a giant conifer that is centuries old. People struggling to deal with emotional problems—whether stemming from dark events of the past, an unlucky roll of the genetic dice, or some other burden—embody the same strength and fragility. Trees provide solace and the assurance that we are not alone in our struggles.

THE BELL OF HOPE

In the same way that healthcare providers look after human health, ecologists and environmental scientists look after the health of the biosphere. One veteran environmentalist is Chris Maser, who helped establish the field of forest sustainability. Maser stated that "the very process of restoring the land to health is the way we become attuned to nature and, through nature, with ourselves. Restoration forestry, therefore, is both the means and the end, for as we learn how to restore the forest, we heal the forest, and as we heal the forest, we heal ourselves."

Whether we wish to help in the recovery of a fellow human being or the restoration of our planet, certain steps are necessary. We must (1) recognize that some dysfunction exists, (2) diagnose the problem, (3) administer therapy, and (4) cease the activities that resulted in the dysfunction in the first place. Of those steps, it is easiest to carry out the first three—notice the shortness of breath, say, and insist on a checkup, where a heart problem is diagnosed and medicine prescribed; or become aware of particulate pollution, trace it to a coal-burning power plant, and see that scrubbers are installed. The final step—prevention—is more difficult. For a human, heart disease generally requires significant behavioral changes: losing weight, for example, or quitting smoking. Reducing non-point particulate pollution can mean expensive retrofitting, which raises manufacturing costs and product prices. Both prevention programs involve considerable personal and social commitment.

There are so many links between trees and human health—trees seen from hospital windows, trees growing in urban parks and streets, trees giving strength and hope to patients in cancer wards. To explain these connections, we would have to invite one last person to our trees-and-healing dinner party: Harvard biologist E. O. Wilson. He could then place his "biophilia hypothesis" on the table as a framework for us. Wilson believes that humans have an innate (or at least extremely ancient) emotional affiliation to other living organisms, and our continued divorce from it has led to the loss of not only "a vast intellectual legacy born of intimacy [with nature]" but, indeed, our very sanity. The health of the planet and that of all its inhabitants, including trees and humans, are intertwined. It is a matter of heart and soul, but also one of action—as a poem by the Bengali writer Vijaya Mukhopadhyay makes clear. The poem's narrator is a tree—a tree that wants to move.

Wanting to Move

Continually, a bell rings in my heart.
I was supposed to go somewhere, to some other place,
Tense from the long wait
"Where do you go, will you take me
With you, on your horses, down the river, with the flames of your
torches?"

They burst out laughing.
"A tree wanting to move from place to place?"
Startled, I look at myself—
A tree, wanting to move from place to place, a tree
Wanting to move? Am I then—
Born here, to die here
Even die here?
Who rings the bell, then, inside my heart?
Who tells me to go, inside my heart?
Who agitates me, continually, inside my heart?

It inspires me to think that just as this tree will somehow follow the bell in its heart, humans will find solutions to the problems of their own health and the health of the biosphere, problems that can seem overwhelming, intractable, and rooted in immobility. Each of us—even a tree—can move from illness to health, from anguish to recovery, and from stasis to freedom.

Chapter Five

PLAY AND IMAGINATION

> Play is the taproot from which original art springs. It is the raw stuff that the artist channels with all his learning and technique.
>
> —*Stephen Nachmanovitch*, Free Play

One of the rewards of climbing into tall tropical treetops is being able to watch animals play. During my twenty-five years of canopy studies in Monteverde, Costa Rica, I have come to feel like a naturalized member of my ten-acre study site's arboreal neighborhood. My initiation into the cloud forest's interspecies community came in 1984, when my colleague Teri Matelson and I documented how birds use epiphytes, the plants that live in the crowns of trees. Ground-based ornithologists who study bird behavior in rainforests are typically limited to peering up at the canopy from a hundred or more feet below. The behaviors of these fast-moving creatures are almost completely obscured by layers of leaves and branches, leaving condor-sized holes in our knowledge of bird behavior and ecology. Teri and I, however, used mountain-climbing gear to scoot up to our treetop perches before dawn and settle in to observe the secret lives of birds. For three months we spent six hours each day aloft, perched quietly on a collapsible cot suspended by slender cords of nylon webbing. When a bird flew into our study plot, we recorded which tree it occupied, what it ate, and how it behaved. Ultimately we had collected more than four thousand observations on over seventy species of birds. We found that over one-third of the visits that involved the collection of food (fruits, seeds, and nectar from flowers) were to epiphytes specifically, rather than the host tree more generally. We also learned that nine species of birds, including the tiny white-throated mountain-gem hummingbird and the seldom-seen spotted barbtail, specialized on canopy-dwelling plants, using epiphytes for over 90 percent of their visits. Those long hours in the air taught us some of the subtle dynamics of the canopy, such as what time flowers would open and when the mist would gather in the forest coves. And they introduced us to the master gamboler of the Costa Rican treetops, the howler monkey.

Every day at 2:30, we would hear the first distinctive calls of the local troop of howlers. Their cries became increasingly loud; soon, the branches of the neighboring trees were rustling, until finally the group sat in pairs and bunches on the limbs that encircled us. They lounged about and groomed one another, but they were also interested in Teri and me. As they surveyed us, I got a creeping sense that we human scientists, perched on our cot and tied in with ropes, were their version of the Discovery Channel—something to hoot over each afternoon. When they tired of watching us watching them, they had other ways of enjoying the canopy. Although some of their behavior involved the serious business of harvesting fruit, vocalizing, and tending to their young, they also simply played. Using their enviable grasping skills, they navigated the three-dimensional world of the canopy as gracefully as a champion skier whips down a slalom course. Sitting on my branch, attached with harness and safety lines, carabiners clanking, I envied their freedom to careen through the canopy, swinging, calling, chasing. It all looked and sounded like the swing set at the primary school where my own children swung, called, chased.

CAPERING IN THE CANOPY

When the green woods laugh with the voice of joy,
And the dimpling stream runs laughing by;
When the air does laugh with our merry wit,
And the green hill laughs with the noise of it;

> When the painted birds laugh in the shade,
> Where our table with cherries and nuts is spread:
> Come live, and be merry, and join with me,
> To sing the sweet chorus of "Ha, ha, he!"
>
> —*William Blake, "Laughing Song"*

The fourth step on the modified Maslow pyramid comprises play, imagination, and free-spiritedness. Without these qualities, our lives would be diminished.

Many behavioral psychologists believe that social play is an evolved behavior that improves the motor and cognitive skills of the young of a wide range of animals. This can yield payoffs in their ability to hunt, forage, and socialize throughout their lives. Howler monkeys and other arboreal primates are particularly well adapted to playing in the canopy. Paleoanthropologists hypothesize that because they evolved in trees, primates developed opposable thumbs, tails, and feet, rotary shoulder joints and wrist joints, and stereoscopic vision specifically for grasping, hanging, and swinging from branches.

Although humans have evolved away from swinging on branches as a means of existence, we still need to play, just like howler monkeys. Through play, children develop physically, learn the ins and outs of social relationships, and cultivate an understanding of the world at large. Even those who are too young to climb high into the treetops find delight by hurtling skyward: the humble tire swing can launch children into space, giving them a sense of movement and freedom that is hard to find while ground bound. Trees continue to play a role in much of our play and recreation, from tree houses, to wooden bats and gym floors, to paths through the forest.

One of my favorite modes of play even today is climbing trees. The type of tree climbing I do now, though, is very different from the grab-and-scramble methods I used as a child to scale the maple trees in my front yard. To climb tall trees requires, first of all, some specialized equipment: a safety helmet and seat harness, inch-wide nylon webbing, carabiners (strong metal hoops to attach the webbing to the harness), and ratcheting mechanical ascending clamps called Jumars. Arborists and beginning tree climbers tend to use static line, which is stiff and has little give to it. I prefer the dynamic ropes that rock climbers use, which have built-in elasticity to take falls. They give me a pleasant "boing" as I climb, and if I get the timing just right, that rhythmic kick helps me move upward more efficiently. I always carry extra webbing and short pieces of rope to tie myself in and secure my equipment. Ropes last for years if cared for properly. Ever since I found two halves of a rope on the forest floor that I had left up in a cloud forest tree overnight—with chew marks at the ends—I have removed climbing ropes each afternoon, replacing them with a tougher, more chew-resistant parachute cord as a placeholder so I need not reshoot the line. Exposure to the ultraviolet rays of the sun can also damage ropes.

Once I'm outfitted with the gear, the real challenge begins: getting a rope over the first branch. This can take me anywhere from five minutes (very rare!) to several days, depending on the height of the branches, the density of the crown, the coordination of my hand and eye, and the patience of my assistants. Professional arborists working with ornamental trees often use a throw-weight—an eight-ounce bean bag with a loop that attaches to a thin, light cord. A skilled thrower can lob such a line over branches fifty feet high. For tall rainforest trees, however, something more ambitious is needed to reach branches that may be as much as one hundred feet off the ground. I learned how to rig and climb such trees from Don Perry, a pioneer in canopy studies. We first met as graduate students, both of us doing fieldwork in Costa Rica, Don at the University of California, I at the University of Washington. A strapping fellow, he would stride out into the rainforest with a heavy wooden crossbow slung over his shoulder. On the stock of the crossbow, he had mounted a fishing reel loaded with twenty-pound test monofilament fishing cord. With the help of a fishing weight, Don could shoot a line up and over branches as high as 150 feet.

When I started climbing on my own, however, I was bothered by the weight and bulkiness of a crossbow, and I found it difficult to string. I also had several nerve-wracking encounters with customs officials, who doubted that the murderous-looking object in my duffle bag was really a piece of scientific equipment, not a weapon. In 1981, on a canopy research tour of New Zealand, I collaborated with a conservationist and fellow tree-climber, John Innes, to invent an alternative to Perry's crossbow. It also relied on a spinning reel, twenty-pound test line, and a two-ounce fishing weight; but instead of the crossbow we used a slingshot attached to a foot-long aluminum rod. The Master-Caster, as we called it, worked like a charm. It was lightweight and could be disassembled easily. Best of all, it looked relatively harmless.

With whatever propellant—throwline, crossbow, or Master-Caster—I get that first critical line up and over the branch, hoping that it falls easily back to the ground on the other side of the branch without hanging up on the foliage, mats of epiphytes, or other branches. Sometimes, I need to coax it down, inch by slow inch, by "twinging," or vibrating, the line like a bass fiddle string as I pay it out. Once that is over, I tie a thicker nylon "parachute cord" to the end of the fishing line, and reel it up and over the branch. This cord is strong enough to take the weight of the heavy climbing rope, which I haul up, hand over hand. The final step is to tie off one end of the climbing rope to the tree (or an adjacent tree), with the free end—at last—ready for climbing.

Once the climbing rope is established, I step into my seat harness and fasten it snugly around my waist, doubling the belt back for safety. For my legs, I use two connected loops of webbing that join to a single Jumar, one foot to each loop. I attach this set of webbing to one of the Jumar's ascenders. The Jumar is a simple but remarkable invention; it has a movable clamp, with tiny teeth through which the rope is threaded. This allows the Jumar to easily slide up—but not down—until I push a knob when I wish to descend. I then attach a length of webbing between my seat harness and a second Jumar, which I clamp onto the rope above the

Jumar that is attached to my leg loops. I alternate placing my weight in one, and then the other Jumar, up the rope to carry me higher and higher into the canopy. In this way, I can "inchworm" upward at whatever speed I wish, stopping to take a photograph, collect a sample, or register the changing perspective I gain as I leave the forest floor and enter the canopy world. Even when carrying out the most scientific of research, my students and I feel comfortable racing one another up to our study branch—or performing graceful (and hilarious) dance moves as we ascend to the epiphytes we are measuring that day.

Researchers are not the only ones who delight in climbing trees. Long restricted to canopy researchers and arborists, the art of climbing trees has now become the domain of the enthusiastic amateur, with the concomitant establishment of climbing organizations, safety standards, and ethical checklists. In 1983, Peter "Treeman" Jenkins, a retired rock climber turned arborist, created Tree Climbing International, Inc. (TCI), a national organization for recreational tree climbing. Finding that techniques used in rock climbing were inappropriate for climbing living trees, Jenkins and others established courses and safety rules for the growing guild of those who wish to climb trees for the fun of it. I recently visited their website and clicked on the "Forums" section. There, I found topic headers on diverse subjects, including "cool tree climbing videos," finding champion trees in Georgia, comparative values of different brands of suspended tents, and how birds help Ponderosa pine trees grow better. The TCI site also provides a venue for exchanging information on climbing accidents and tree damage. The hope is that the tree-climbing community will learn from the experiences and errors of others by becoming aware of the situations that signal danger or damage to human or tree. Many small companies and individual guides now make a business of training and taking people up into the treetops, offering adventurous types an "up and down" climbing experience and, for the brave-hearted, the opportunity to move from one tree to another in mid-air.

Some canopy researchers and arborists have concerns about all this climbing, especially where companies take customers repeatedly to the same group of trees. What are the impacts on the trees of soil compaction, bark rubbing, and disturbance of canopy-dwelling mosses? Are those who are most interested in trees imposing harm? Few long-term studies address these questions, since the reaction time of trees is slow compared to the short time recreational climbing has been part of the adventure sports industry. Concerned citizens of the canopy world have initiated informal monitoring projects to record and report potential harm to trees and to develop ways to mitigate those effects.

DREAMING ALOFT

To many climbers, the ultimate experience is spending a night high in the treetops. There is something about sleeping in the forest—whether on the ground or in the trees—that brings us as close as we can get to nature.

My first overnight experience suspended in a hammock between branches of a giant tropical rainforest tree remains vivid in my memory even thirty years later. I climbed into the canopy as the sun set, the darkening understory giving way to the lighter environment of the canopy—though that, too, gradually became part of the jungle night. Bird songs gradually yielded to the buzzing, whirring, creaking calls of insects, which grew louder both below and above me, a sort of stereo effect I had never heard before. I curled up on my hanging cot, water bottle and a bag of snacks tied to an auxiliary cord, my harness and rope giving me a sense of security as the spookiness of being two hundred feet above the ground crept into me. At some point during the night, an anteater rambled over to my perch, in pursuit not of a dormant human but rather of the steady stream of leaf-cutter ants that were harvesting chunks of leaves from the trees and walking them along the branch highways down to their underground nests. On seeing me, the collie-sized mammal seemed as startled as I had been. But we looked at each other for a long moment without fear, two arboreal animals in a high place on a dark night.

Since that time, I have spent many nights aloft. What has surprised me is not the "otherness" of the canopy night compared to where we ground-bound humans normally sleep, but rather how homey and comfortable it seems up there with darkness stretching out in three dimensions. We were raised with the classic lullaby, "Rock-a-bye baby in the treetops," with its inevitable and sobering conclusion: "and down will come baby, cradle and all." And there are noxious insects and poisonous reptiles somewhere up there. But during those nights I spent on my canopy cot, swaying slightly in the wind one hundred feet above the ground, I couldn't have felt safer and more ready for sleep, lulled by my nocturnal companions above, below, and around me.

GAMES THAT GO WITH THE GRAIN

Take me out to the ball game . . .

—*1908 Tin Pan Alley song by Jack Norworth and Albert Von Tilzer*

A sport that is inextricably linked to wood—and therefore trees—is our national pastime, baseball. When baseball was invented in the 1850s, bats, which came in all shapes and sizes, were made of hickory, an extremely hard and heavy wood. The hickory sticks wielded by Babe Ruth, for example, weighed up to 47 ounces—almost three pounds. Although that gave him a hefty "sweet spot" for hitting home runs, today's major leaguers prefer bats that weigh no more than 32 ounces, allowing for more time to analyze the pitch, quicker swings, and greater control.

Although major league rules don't regulate weight, they do specify that a bat must be no more than 2¾ inches in diameter and no more than 42 inches long, and must consist of a single piece of solid wood. Typically, bats are fashioned from white ash or sugar maple that is harvested from deciduous forests in Pennsylvania and New York from trees that are around fifty years old. These woods are loved for their hardness, durability, strength, and "feel."

Metal bats, introduced in the 1970s in amateur baseball as a cost-saving alternative to breakable wooden bats, proved to be more efficient than wood, in one study clocking average ball speeds of 103.3 miles per hour, versus 98.6 mph for wooden bats (leading some athletic associations to ban the use of metal bats in high school play for safety reasons). Considerations of both safety and tradition have kept metal bats out of professional baseball. And even fans of metal bats agree that only wood delivers the satisfying "crack" that we associate with a well-hit line drive.

The history of racquet sports, too, is intertwined with trees. As early as the fourteenth century, people used a wooden, gut-stringed frame to bat balls around. In 1967, however, Wilson Sporting Goods introduced the first metal tennis racquet, the T2000. Stronger and lighter than wood, it became a top seller. Later, other materials were used, including graphite (carbon fibers bound together by a plastic resin), ceramic, fiberglass, titanium, and Kevlar. Sports such as squash and badminton soon followed with metal and synthetic racquets of their own. Wood no longer offered anything that other materials couldn't provide better—except for collectible value and aesthetic appeal.

One of the few pieces of sports equipment that has not been replaced with synthetic materials is the pool cue. Consider the archetypal image of the pool shark, scoping with hooded eyes his next shot through a haze of cigarette smoke. The cue he cradles in his hands is made of straight-grained maple from the northeastern United States. Snooker cues, in contrast, are made of ash. Both of these woods provide players with the same important characteristics of resilience, strength, and straight grain that they provide to the baseball player at bat.

Golf is a sport I associate with trees in a negative way, because it has led to the creation of monospecific grass swards, often where woodland used to be. Even though they look healthy and green, golf courses generate runoff that is high in nitrates and pesticide residues—more ecological bad news. Nevertheless, I must note that two of golf's primary tools are contributed by trees. The clubs known as the woods are the longest in the bag and are used mostly for distance shots. In Scotland, the earliest woods were made by bow makers and featured heads carved out of beech, ash, holly, dogwood, pear, and apple. Later, the heads were made of persimmon or maple. Today they are chiefly made of steel or a titanium alloy, though beautiful specialty persimmon clubs are still manufactured. Shafts have also undergone a metamorphosis. Originally they were made of ash or hazel to give the club whip; however, the arrival of golf in America in the early 1800s led to the use of hickory in the shafts, which is far stronger than most other woods. Hickory remained the standard until steel shafts were introduced in 1925.

The other golf-related material that came from trees is, surprisingly, part of the ball itself. The first balls were leather-covered objects stuffed with boiled goose or chicken feathers. These fragile "featheries" were used for almost four centuries—until 1848, when the "gutta-percha" ball was created. Gutta-percha is the evaporated milky latex from the *Palaquium gutta* tree of eastern Asia. This and related trees, such as the "bully tree" of the West Indies, yield a hard, rubberlike material called "balata." When zigzag gashes are cut in the bark, the latex can be collected in cups; it then turns into a tough, resilient, and water-resistant material. Here again, though, the "gutty" went out of use, around 1915; today, the core of golf balls is made of Surlyn, a commercial thermoplastic polymer derived from petroleum.

Not all sports evolved for sheer amusement. Archery is a recreational activity that evolved out of necessity. As stone arrowheads discovered in Africa tell us, it existed as early as 25,000 B.C. as a means of procuring food. Since then, bows and arrows have been used in warfare and for sport worldwide. Early bows were crafted from a single piece of wood with a strand of sinew strung between the tips. In Britain, the wood of the strong and flexible yew tree was used most frequently, giving rise to the name "yeomen" for the archers of ancient England. From the thirteenth to the sixteenth century, the amount of yew wood needed by Britain for war archery exceeded domestic sources. After all of the yew stands in the British Isles were depleted, the government began to import the wood from Spain. During the 1500s, Bavaria and Austria exported nearly one million yew staves; by 1560, not a single yew tree remained in Bavaria. In 1595, Elizabeth I decreed that firearms were to replace military longbows, not because firearms were superior but because no usable yew wood was left in all of Europe. Even at the battle of Waterloo almost two hundred years later, firearms still could not match the lightning speed and precision of the yew longbow. Today, the few ancient yew trees remaining in Great Britain are protected in churchyards.

Modern archery bows consist of wood laminated with resins, fiberglass, plastic, and epoxy. And arrows, which were originally crafted from wood and eventually evolved to incorporate tips made from stone and metal, with feathers for aerodynamics, today are constructed from an aluminum alloy or fiberglass, with steel points and plastic fins.

Guns, also often used today for recreation, grew out of the unfortunate necessities of war as well. Like bats and rackets, guns have a strong connection to trees—as I learned from Len, a longtime member of the Capitol City Rifle and Pistol Club in Littlerock, Washington. The best wood for solid wood stocks, he told me, is black walnut, which is hard but not brittle, dense but not heavy, and has an attractive grain. Len showed me one of his choice rifles, and even I found the black walnut gunstock to be a thing of beauty. Walnut is also stiff, absorbing the force of recoil. However, because it is in high demand for the furniture industry, it is often out of the price range of gun aficionados. Cheaper wooden stocks are available as a result, made from other hardwoods, primarily beech, a light-colored wood with little grain or character, but one that can be stained to resemble walnut. Birch, maple, myrtle, and mahogany are also used. Laminated stocks, made of layers of hardwood glued together under pressure, are the strongest. Len reluctantly admitted that synthetic stocks, consisting of an injection-molded fiberglass shell filled with plastic foam, have some advantages over wood. Elaborate and extensive patterns of any color can be included at little expense, for example. Synthetic stocks also flex to moderate the effect of heavy recoil. However, Len said, synthetic stocks can be *too* flexible, and they can warp in hot weather and freeze in cold weather, becoming so brittle that they shatter. Despite the higher cost, Len sticks to wood over synthetics "because it just feels right in my arms."

Other sports began out of utility as well. Pole-vaulting originated in the marshy provinces of the Netherlands to enable farmers to jump over canals and brooks. Modern competitions began around 1850. A vaulting pole is a complex thing, as it must be able to absorb all of the vaulter's energy while bending, and then return that energy as it straightens out. Initially, vaulting poles were stiff, made from wood or bamboo with a sharp point on the bottom. Later, new materials led to lighter, more flexible vaulting poles. Today, carbon fiber is the preferred material.

Skiing grew out of the need to navigate terrains of snow and ice. Early skis preserved in the peat bogs of Scandinavia have been dated by pollen analysis to the year 2500 B.C. These paddlelike skis, made of single pieces of carved wood with an upturned tip, were likely used to track herds of game. In the mid–nineteenth century, the predecessor of the modern downhill ski was created in Telemark, Norway. Its design and the new binding—made of twisted tree roots, rather than less flexible leather—allowed the ski to turn when it was set on edge and gave the skier unprecedented control on downhill runs.

Although skis were essential for survival, they were soon being used for recreation. In the 1860s, to break the monotony of long winters, mining camp residents in California competed in downhill races. The first match-ups were informal affairs, but soon they became hotly contested regional events. Equipped with twelve-foot-long skis made from boards of hickory or ash, miners from different camps raced one another on courses that plummeted straight down open mountainsides. Later, they developed skis that were grooved on the bottom; applying wax to the boards, they achieved speeds in excess of ninety miles per hour. As downhill skiing grew in popularity, so did the demand for more reliable equipment with better performance. Currently, ski manufacturers use wood or synthetic cores or composites of wood and injected foams. One company in

Washington State still uses locally grown spruce and fir for its cores. Spruce, being a long-fiber wood, is lightweight and flexible, while fir is denser and has greater strength and resilience, so a blend produces a fine balance. Most modern skis consist of corrugated wood channels sandwiched between laminated wood layers to reduce weight while maintaining strength.

Waterskiing was invented in 1922, when eighteen-year-old Ralph Samuelson of Minnesota asked, If you can ski on snow, why not on water? First young Samuelson tried barrel staves, then snow skis, and finally he fashioned the first dedicated water skis from lumber he shaped in his father's workshop. Samuelson made his bindings from leather strips and used a long window sash as a ski rope. Modern water skis are made of carbon fiber, fiberglass, titanium, and aluminum, in addition to wood. However, a considerable proportion of enthusiasts still choose wood, linking lake to forest with the graceful parentheses they create in the wake of the boats that pull them through the water.

What about "skiing" on pavement—with the assist of wheels? After a new skateboard park was installed in my town, I occasionally stopped to watch sidewalk surfers, with deceptive casualness, do their amazing swoop-flights on the scooped-out concrete deck of the park. Last year, I tried skateboarding myself, getting up early to avoid collisions with the experienced boarders who practiced their art after school. My own teenaged children urged me to put a bag over my head so that I wouldn't be recognized by their friends. I managed to learn the basics, but the elegance and freewheeling zooms of the regulars have remained elusive.

I did enjoy telling the silent youths with whom I shared the park—to no visible response—that their sport is inextricably linked to trees. A skateboard has three parts: the board (or deck), the wheels, and the trucks, which connect the wheels to the board and allow the board to turn. Modern skateboards are made from seven plies of sugar maple veneer, pressed together using polyvinyl glues at a pressure of about three hundred pounds per square inch. After days of curing, the makers, using hand routers, cut out the final shape, apply edge trimming, paint the board, and send it on its way. Though it may seem "low-tech," a laminated sugar maple deck is superior to alternate materials such as epoxy, fiberglass, and carbon-loaded thermoplastic nylon in terms of toughness, elasticity, and feel. An extra finishing layer of Hawaiian koa, bamboo, or rosewood adds still more strength and creates a more responsive ride.

Another area where durability underfoot is required is in the floor on which indoor sports such as basketball are played. Although the floor gets far less attention than the athleticism displayed on the court, it is a critical part of the game, contributing to safety and player performance. A sports floor must be durable and resilient in the face of the pounding of heavy athletes, and it must provide just the right amount of friction to prevent falls yet allow for slides and quick turns. As a player's foot hits a sports surface, part of the force is absorbed by the floor, and the rest is returned to the athlete. Some 150 years ago, northern hard maple, or sugar maple, became the sports floor of choice because of its elasticity and traction. It is close grained, hard fibered, and free from slivering and splintering, characteristics that help reduce the incidence of injuries. For example, one study was performed on women athletes who sustained knee ligament injuries while playing handball. The incidence of this severe injury was 0.82 injuries per 1,000 playing hours for those playing on synthetic surfaces, contrasting with a much lower risk for those playing on wooden floors. In 1999, nearly 23 million square feet of maple sports flooring was installed in the United States, a little more than half of flooring of all types. This is equivalent to about 115,000 trees seventy feet tall and fourteen inches in diameter.

On the lighter side of athleticism, let's look at conkers, a now rather obscure child's game using the nuts of the horse-chestnut tree. The game is British, but the name comes from the French word *conque*, meaning "conch," as the game was originally played using snail shells. The horse-chestnut tree, native to the Balkan Peninsula, has been grown in Britain since the seventeenth century. Trees make large fruits with a spiky covering that breaks into three parts, releasing the shiny nuts. The game is played by schoolchildren (mostly boys), who thread the nuts on strings and suspend them, singly, from the hand. The goal is to break the opponent's conker with a single blow. The contestants take turns swinging until one of the conkers is cracked or smashed. Speed and accuracy, as well as conker strength, are key. Scoring involves keeping count of how many wins an individual conker makes—a two-er, a five-er, etc. Each year, the World Conker Championships take place on the second Sunday of October in Ashton, Northamptonshire, England. In 2004, an audience of five thousand turned up to watch more than four hundred international competitors contest the title, which currently stands at a 5000er. Despite the apparent frivolity, proceeds from the event go to the purchase of "talking books" for the blind.

Another prominent wooden contributor to fun and frolic is the roller coaster, which started out as a slide structure in Russia around 1600. Today, although most modern roller coasters are constructed from steel, many wooden ones remain in business. According to aficionados, "steel can be cold, but wood flexes, creaks, rattles, fishtails, and gives you more of a feeling of being a part of the ride." The largest roller coaster in America is the "Son of Beast" at Paramount's Kings Island in Ohio. This ride, which attains maximum speeds of eighty miles per hour, required 1.65 million board feet of lumber (made from southern yellow pine and Douglas-fir), over 50,000 pounds of nails, and 225,000 bolts for its construction. To maintain safety, crews climb the structure every day to tighten the bolts. Over 214 feet at its highest, "Son of Beast" is the tallest roller coaster ever built and one of the biggest wooden structures in the world.

In the cacophony of these sticks, poles, bats, floors, and thrill rides, let's also remember that trees as forest members can become elements in the game. One summer when I was in college I worked as a forest surveyor for the U.S. Forest Service in southeastern Alaska. My co-workers and bunkmates were snuff-dipping, big-bearded men who made their living cutting the huge hemlock, spruce, and cedar trees of the temperate rainforest. They felled trees, set chokers (heavy metal chains that are

20. Loggers are dedicated to both the work and play involved with harvesting trees from the forest. Here, they compete for prizes that reward prowess in the skills used for their work. Reproduced by permission of the Department of Forest Products, University of Idaho College of Natural Resources.

looped around huge logs, then pulled up by the "tower" at the landing), and danced up and down steep clear-cut mountainsides by balancing on felled trees. On the fourth of July, all of us flew into the small town of Ketchikan to enjoy a parade and fireworks. Play trumped work on that day. The highlight was the Logger Sports Competition, with contests in axe throwing, choker setting, and spar climbing using pole spikes and a leather belt, all carried out in a friendly spirit of rivalry and fun.

Another sport that involves both whole trees and muscled participants is the caber toss, a traditional Scottish athletic activity. Some trace the origins to the Gaelic word for rafter or pole, or a shortened version of "casting the bar." Others say this test of strength, balance, and agility came from foresters practicing the laying of logs across creeks to make bridges, or from soldiers tossing logs across moats during castle sieges. In this sporting event, the kilted participants literally toss a large wooden pole, or caber, head over heels, across flat ground. There are no standards as to a caber's length, weight, or type of wood, but traditionally it is sixteen to twenty feet long and weighs 80–130 pounds, with the record an astounding 280 pounds. Amazingly, even though few objects are more awkward to maneuver than a log, the caber is not thrown for distance but for style, which includes stance and steps, balance, and agility. The athlete lifts the caber, runs forward to gain enough momentum to head the pole in the proper direction, and then heaves it up so that it arcs, flips, and lands with the end pointing directly across from the tosser. Like a lucky bowling ball in an amateur bowling league, some cabers are brought out for games year after year, a particular tree bringing pride to the clan of the tosser across generations.

For the past thirty years, tree climbers and arborists have had a special event of their own: the International Tree Climbing Championship. The founder, Dick Alvarez, is a Californian who wished to reinforce the skills that allow an arborist to perform his work safely. His annual workshop grew into a full-blown jamboree that now moves from city to city and features many contests related to the profession: the work climb, aerial rescue, rope throw, and body thrust speed climb, to name a few. The championship, which in 2006 brought together more than a thousand contestants, has expanded to include Europeans, Canadians, and separate women's events since its modest beginnings in 1976.

Finally, there is the quiet sport of hiking, in which we humans use trees not as sticks or sliding surfaces or structures, but simply as themselves. When I wake up each morning, the first thing I see is the shelf that holds my hiking books, thirty-nine brightly colored volumes that describe trails in the Pacific Northwest. Just seeing their spines, with titles like *Loop Trails of Washington*, *Fifty Trail Runs in the Pacific Northwest*, and *Best Hikes with Children in the Cascades*, jogs memories as well as anticipation for future wilderness journeys. Each is slotted with bookmarks, indicating possible trips with friends or for going solo into the forest to find beauty, renewal, silence, and play—and to visit something of the wild within myself.

ARBOREAL ARTS

The bonsai tree
in the attractive pot
could have grown eighty feet tall
on the side of a mountain
till split by lightning.
But a gardener
carefully pruned it.
It is nine inches high.
Every day as he
whittles back the branches
the gardener croons,
It is your nature
to be small and cozy,
domestic and weak;
how lucky, little tree,
to have a pot to grow in.
With living creatures
one must begin very early
to dwarf their growth:
the bound feet,
the crippled brain,
the hair in curlers,
the hands you
love to touch.

—Marge Piercy, "A Work of Artifice"

Many hobbies entwine humans with trees, in ways that modify both. One of the most notable of these is bonsai, a living art form practiced by diverse cultures. In the Japanese language, *bonsai* means "tray planting," the art of dwarfing trees and developing them into appealing shapes by training them in containers. Every branch and twig of a bonsai is shaped until a particular image is achieved, though the artist strives to betray no human interference. The art form was first cultivated in China over a thousand years ago, where the practice of growing trees in pots resulted in miniature trees that displayed sparse foliage and rugged, gnarled trunks. It was introduced to Japan during the Kamakura period (1185–1333) via Zen Buddhism, which was then spreading throughout Asia. The reduction of things to their essential elements was an important aspect of this philosophy, and bonsai manifested that doctrine beautifully. In the mid–nineteenth century, after more than two hundred years of isolation, Japan opened up to the rest of the world. Travelers who visited Japan described miniature trees that mimicked mature trees in nature, and subsequent exhibitions in London and Paris opened the world's eyes to bonsai.

Bonsai are ordinary trees or plants, not special hybrid dwarfs. They develop from seeds, cuttings, or young trees, and at maturity range in height from two inches to three feet. Bonsai are trained by pruning branches and roots, pinching off new growth, and wiring the branches and trunk. Pine, azalea, camellia, bamboo, and plum are often used. Bonsai can live for hundreds of years, with specimens passing from generation to generation.

The first time I saw bonsai trees was during a fifth-grade visit to the National Botanical Gardens in Washington, D.C. I remember feeling two strong and conflicting emotions. Part of me was fascinated by the tinyness of what is usually gigantic—a huge tree reduced to the size of a table lamp, but with all the structural elements intact. Another part of me was horrified by the dwarfing and imprisonment of what should be grand and free, manifesting our conceit that we can conquer nature. The fascinated side of me wanted to invite some tiny Japanese dolls and have a pretend tea party beneath those perfectly cast branches; the horrified side wanted to release the tree's weights, grab the pot, and carry it outside to liberate the tree so it could grow into its real form.

Topiary, the art of sculpting shrubs and trees into recognizable shapes, is an outdoor and enlarged version of bonsai. The word derives from *topiarius*, Latin for an ornamental landscape gardener, and the practice—

not surprisingly—dates from Roman times. Pliny the Younger, who occupied a villa in Tuscany with elaborate grounds, described the elaborate animals and obelisks fashioned in clipped greens at his Tuscan villa. One part of his garden was taken up with shrubs that had been shaped to spell the names of Pliny and the topiary artist. Revived in Europe in the sixteenth century, topiary is associated with the gardens of the European elite. Combinations of abstract styles and traditional geometric shapes now enhance many gardens on both sides of the Atlantic. The plants used in clipped topiary are evergreen and have small leaves or needles and compact growth habits. Common plants include boxwood, bay laurel, myrtle, yew, and privet. Another method for creating topiary is to plant climbing vines at the base of a hollow or moss-filled wire frame and to train them over and into the frame, a three- to six-month-long process. Maintenance consists of monthly feeding with plant food and fungicide, and monthly trimmings to keep plants in form.

There is another way of "working" trees, one that we sometimes see in the natural world in old orchards or urban parks: cases of self-grafting, where a branch grows back into itself or into another tree to form a living union. Certain tree species—the so-called inosculate species, which include fruit trees, sycamores, elms, and willows—do this readily. A few gardeners of vision, patience, and humor have used this capacity to engage in what they term "arborsculpture." The grand old man of arborsculpture, Axel Erlandson (1884–1964), was inspired by observing a natural graft between two sycamore trees. He began to shape trees by planting them in patterns and then pruning, bending, and grafting them. What began as a hobby soon grew into his passion and supplied a meager but adequate livelihood. He created the "Tree Circus" in his own yard in Santa Clara, California, which opened for paying visitors in 1947 and intrigued his guests for many years. Today, these living sculptures are on display at Gilroy Gardens (formerly Bonfante Gardens), a theme park attracting wonder side by side with giant thrill rides bearing such names as "The Demon," "Top Gun," and "The Vortex."

Woodcarving, a hobby that combines trees, art, and functionality, has occupied the hands and spirits of humans since they learned how to sharpen bits of flint around a fire. Woodcarving ranges from decorative bas-relief on small objects to life-sized three-dimensional figures, furniture, and architectural decorations. Wood is not the easiest material to work because of its tendency to crack, its vulnerability to insect damage, and its proclivity to twist with changes in humidity and air temperature. I learned a bit about woodcarving in 1978, when I lived for six months in France. I took a class in woodcarving in the 13th Arrondissement of Paris from Monsieur LeBec, an older gentleman with a goatee and black beret. When I arrived for my first lesson, his students and apprentices were all hunched silently over slabs of wood, carving bas-relief oak leaves within a square border, each piece exactly the same. It hardly seemed the creative activity I had envisioned. Wanting to "liberate" a figure from a chunk of wood, as Michelangelo had done with marble, I informed M. LeBec that I should like to carve a snail. "Un escargot?" he asked askance, raising one eyebrow. He sighed and gave me a piece of oak wood that I chipped away at until I had produced a creature that could pass, with some imagination, for "un escargot." Although it was neither zoologically correct nor particularly aesthetic, I had a fabulous time carving it, and it now rests on our mantelpiece, a French snail, slightly askew, holding in its grain the satisfying hours I spent creating it.

The subject of wooden objects and their role in play is not complete without considering wooden toys for children. Toys, of course, are an important part of how we learn about the world. Playing with toys helps youngsters to learn cause and effect, explore their creativity, practice skills they will need as adults, and discover their identity. The earliest toys—dolls representing infants, animals, and soldiers—were made from materials found in nature, such as rocks, clay, and wood. Today, thousands of companies create and market wooden blocks and boats, puppets and puzzles. When my daughter, Erika, was young, she was given a doll made out of a single piece of wood and dressed in bits of cloth and ribbon. Her name was "Mercilee," and she accompanied us on trips near and far. After many months, her pieces of fabric wore away, and the grain of the wood emerged with surprising beauty: she became a simple piece of smooth, curved wood, her worn contours like the soft skin of Erika. Today she rests on our mantelpiece—right beside the escargot.

REACQUAINTING OUR CHILDREN WITH TREES

One of the joys of being a forest ecology professor is that I'm frequently asked to give talks about trees and nature to the public. I especially love to give ranger-style campfire talks at regional and state parks. When adults and children gather around the fire to listen to my spiel and ask questions, or to share their own stories about encounters with wildlands, I feel reassured that despite the pervasiveness of television, video games, and the Internet, the forest is still perceived as a place of renewal and re-creation.

On a recent peaceful Sunday morning, my husband, Jack, made thimbleberry pancakes, and our neighbors came over to eat them with huckleberry jam that Erika and her friend had made the day before, after plundering every bush for acres around. I felt deep satisfaction as I watched the kids zoom into the woods, knowing that their feet were learning the paths of their home forest, just as mine had in my own childhood in woodlots around our house. But recent trends indicate that young people are losing contact with nature at an alarming rate. Instead of passing summer months hiking and telling stories around the campfire, children are more likely to attend computer camps or weight-loss workshops. Nature has become more an abstraction than a reality. Children now spend an average of forty-four hours a week pursuing activities that have nothing to do with nature, from plugging into electronic media to participating in organized sports on tamed playing fields—and then there are the hours they spend sitting in the car, being transported from one activity to the next. After tens of thousands of years of playing outdoors, kids today are couched far too comfortably away from nature.

Journalist Richard Louv is very concerned with these patterns. He spent ten years traveling around the country speaking with parents, children, teachers, scientists, religious leaders, child-development researchers, and environmentalists in rural and urban areas about their experiences in nature. In his book *Last Child in the Woods: Saving Our Children from Nature-Deficit Disorder*, he reported that children have become increasingly alienated from the natural world, with ominous implications for not only their physical fitness, but also their mental and spiritual health. Children who receive early positive exposure to nature, he suggested, thrive in intellectual, spiritual, and physical ways that their "shut-in" peers do not. Although advances in technology provide young people with an astonishing range of information and virtual experiences, these advances have also made it far easier for children to spend less time outside.

Why is this happening? At first Louv thought that dwindling access to nature because of urban and suburban development was keeping kids from going outside. That would seem to be a factor in my own childhood neighborhood in Bethesda, Maryland, which when I was young had many forest patches and fields where my siblings and I made forts, played hide-and-seek, and collected fireflies on summer evenings, and where today I see only large homes standing shoulder to shoulder on well-curated lawns. However, Louv found that even where children have access to wildlands, they rarely venture outside. One explanation for this trend is parents' anxious attitudes about the world—what Louv calls "stranger danger," a nebulous fear about violent criminals and sexual predators, traffic accidents, and lawsuits. Ironically, according to the Duke University Child Well-Being Index, violence against children dropped nearly 40 percent between 1975 and 2003, but media coverage has grown more shrill, literally "scaring children straight out of the woods." Children who have not had sufficient exposure to nature, Louv says, may fall victim to what he calls "nature-deficit disorder," which "can occur in individuals, families, and communities" and leads to "diminished use of the senses, attention difficulties, and higher rates of physical and emotional illnesses." In fact, many studies show a relationship between lack of parks and open space, on the one hand, and such urban maladies as high crime rates and depression, on the other.

Louv has clearly struck a nerve. His book has sold very well, and he has been interviewed hundreds of times in the media. He attributes this response to a profound desire to reintegrate our children in nature. Happily, he believes this is possible. Nature play could, he suggested, emerge as a promising therapy for attention-deficit disorder. He also sees the emergence of many nature-centered children's activities in nontherapeutic contexts: summer camps, school programs, and adventure educational experiences. One such purveyor is IslandWood, a residential environmental learning center on Bainbridge Island in Washington State, which brings fourth-, fifth-, and sixth-graders from urban and rural areas of Puget Sound for four-day experiences on 255 acres of wildland. IslandWood's founder, Debbi Brainerd, recognizes the necessity of children experiencing the trees, birds, streams, mosses, and slugs of their native biota. The kids themselves, after a bit of initial hesitancy, jump into the activities with enthusiasm. They link nature with fun, but more critically, they engage with wilderness and wildness. As Wallace Stegner concluded years ago, "Something will have gone out of us as a people if we ever let the remaining wilderness be destroyed. We simply need that wild country available to us, even if we never do more than drive to its edge and look in. For it can be a means of reassuring ourselves of our sanity as creatures, a part of the geography of hope."

Fortunately, many dedicated people embrace this notion with passion, taking it into their work as protectors and stewards of the land. In 2005, the Nature Conservancy of Washington conducted a set of interviews with its staff and board members to learn what motivated them to strive, at relatively low pay and over long hours, to "save the last wild places." Surprisingly, rather than citing reasons like an intense interest in biodiversity or bird-watching, almost every individual said that they had had some formative—even transcendent—experience with nature when they were a child. Many described in great detail the experience of sharing nature in the woods or by the ocean with a parent, a scoutmaster, or a family friend. Going fishing with an uncle—experiencing water as an unruly rushing river rather than dripping dully from the bathtub tap, and handling a fish rainbowed and strong instead of seeing it dead and flattened in a supermarket display—led them to a career in conservation. If those experiences are absent in the lives of our own children, displaced by television, video games, and YouTube, where will the next generation of forest stewards come from?

RAPPING IN THE FORESTS

I believe that arousing a concern for trees and nature in young people is one of the most critical challenges we face in the coming decades. However, as a middle-aged, middle-class female scientist, I have little hope of capturing the imagination or even the interest of most young people, and especially not of urban youths, whose experiences, values, and knowledge sets are so different from my own. One strategy I have devised to overcome that distance is to work with intermediaries: people who connect with youths, and who have a fascination with nature. I found one such person at the Evergreen State College in 2002. George "Duke" Brady is from San Francisco. His great passion is rap music, and his nightlife includes performances at hiphop clubs. He took my forest ecology class as a freshman and became interested in ecosystem ecology during our field trips to the Olympic Peninsula. Once, I took him up to my canopy platform on campus. Delighting in the new environment, he spontaneously spun out a rap song:

Forest Canopy Freestyle Rap

Wet and green moss,
I'm at a loss
to describe the beauty
falling on my booty.
But held up by strings
came up here to do some things.
But no pressure,
I'm feeling free,
I walk on the ground—
but I wish I was a bird.
In the treetops we walk often

but never see the tops.
And I'm coughing up air
into the atmosphere.
But it seems now my hair
is part of everything around here.
I dare say that I love being here
and I could spend a minute of every day here,
like a hermit,
I would never burn it
or chop it down.
Rhyming e's and i's
I don't know why
I feel high up in the sky,
haven't smoked anything,
haven't eaten mushrooms,
but I still want to sing,
I keep opening my eyes
and want to stop and just breathe,
leave it alone.
Take what you bring in,
pack it in pack it out
take what you own,
pack it out pack it in.

When I later gave presentations to school kids and played a recording of Duke's rap song, even the children seemingly uninterested in rainforest snakes, strangler figs, and gliding arboreal mammals came to attention. Rap music could connect urban kids with the forest.

In the summer of 2004, I expanded this strategy into a weeklong project, enlisting the help of two other ecologists at Evergreen (an entomologist and a marine biologist), youth counselors, and a rap singer with the surprising name of C.A.U.T.I.O.N. Our objective was to get forty middle-school children from the city of Tacoma interested in nature by linking it with activities that they find enjoyable and meaningful.

On Monday, C.A.U.T.I.O.N. and I walked with the students to the Evergreen campus forests and I taught him how to ascend into the canopy with climbing gear. As he inched his way up, I told the students that C.A.U.T.I.O.N. would be going where no person had ever gone before, and would be seeing completely different types of plants and insects than we were seeing on the forest floor, even though he was only sixty feet above us. My words helped them realize that, in ecological terms, the canopy microhabitat fosters a disjunct flora and fauna. But the same message hit much closer to home when C.A.U.T.I.O.N. returned to the forest floor. Even before he stepped out of his harness, he had a spontaneous piece of poetry ready to go, and the students gathered around him to hear his words and beat. The next day, the entomologist led an experiment with the thatch-mound ant colonies in the college parking lot, and students learned how to test a hypothesis about homing abilities in the ants—after which, C.A.U.T.I.O.N. was ready with a rap song about evolution. On Wednesday, the marine biologist led students in collecting marine organisms in the intertidal zone, and our singer was there to spin some rhymes with starfish, barnacles, and clamsquirts.

On the afternoons of those three days, the kids broke into small groups, and C.A.U.T.I.O.N. helped them develop musical interpretations of their field experiences. On the last two days, audio engineers recorded the original lyrics they had created. By the end of the week we had produced a CD of nature-inspired, C.A.U.T.I.O.N.-guided music. The kids were hugely proud of the CD and said they were eager to share it with their friends, family, and schoolmates. Each afternoon, after all the work was done, participants formed a "talking circle" and discussed what they had learned from that day's experience. At the final talking circle, most of them commented on how much they had learned about the hidden worlds of trees, insects, and marine animals, and said that they would recommend the program to their friends. The critical ingredient leading to the project's success, however, was C.A.U.T.I.O.N., who in addition to being energetic and artistic, connected with both urban youths and field ecology.

In every community, opportunities exist to traverse cultural and generational divides, and in so doing, to foster the next generation of humans who will draw inspiration from nature. Play is a fundamental human need, and trees provide a bounty of materials and places for it. Swinging from a horizontal tree limb on a golden afternoon, screaming down a wooden roller coaster, or hearing the satisfying smack of ball against wooden bat reminds us that trees help humans step up through the fourth level of Maslow's pyramid and experience the joy expressed by poet e. e. cummings:

> i thank You God for most this amazing
>   day: for the leaping greenly spirits of trees
> and a blue true dream of sky; and for everything
>   which is natural which is infinite which is yes

Chapter Six

CONNECTIONS TO TIME

> When a tree is cut down and reveals its naked death-wound to the sun, one can read its whole history in the luminous, inscribed disk of its trunk: in the rings of its years, its scars, all the struggle, all the suffering, all the sickness, all the happiness and prosperity stand truly written, the narrow years and the luxurious years, the attacks withstood, the storms endured.
>
> —*Hermann Hesse, "On Trees"*

When my family first arrived in Olympia, we made our home on five wooded acres off Steamboat Island Road. Along with some good Douglas-fir trees, a single Pacific yew, and many cedars and hemlocks, the property featured wild rhododendrons, which were in glorious full bloom when we moved in. Over the past fourteen years, we have added five chickens, a trampoline, a weed-dominated vegetable garden, some struggling fruit trees, and a deck with a picnic table made by my father-in-law. When I do the dishes, I look out on an ancient maple tree. That tree tells me the season, just as my watch tells me the time of day.

Trees express time with a precision and beauty that are unmatched in nature. Changes in their foliage mark the passage of Earth's seasons, while the incremental growth in their rings mark Earth's years. Nothing more effectively indicates seasonal transitions than the tender green of the emerging buds of spring, the rich, deep greens of summer, the multicolored leaves of autumn, or the delicate filigree of snow on tiny twigs after a winter storm. We are inspired by trees' relationships to time: the great age they can attain and the fierce disturbances they can endure. When we walk along the winding paths of a cemetery, we pass beneath the trees that have dwelt there far longer than their interred neighbors, giving us a comforting sense of continuity.

COUNTING THE YEARS

> This pine tree by the rock
> Must have memories, too—
> After a thousand years,
> See how its branches
> Lean toward the ground.
>
> —*Ono no Komachi*

In the past, trees epitomized stasis, but recently, scientists have reassessed that notion. Relying on clues as diverse as thirty-foot-long mud cores and microscopic pollen, they have arrived at a new understanding not only of trees, but also of time and history—the fifth level of Maslow's modified pyramid.

To understand the relationship between trees and time, we must first determine how old they are. With some trees, age can be arrived at merely by looking. Douglas-fir, for instance, has an orderly way of growing, producing a regular number of buds at the tip of each twig each year. The buds lengthen to become the central and side branches during that growing season, after which each branch produces a regular number of buds at its own tip. This pattern of growth results in a regular set of visible whorls, or rings of branches, along branches and trunks, which gives the tree its overall shape as it matures. To estimate the age of a young Douglas-fir tree, therefore, one simply has to add up the number of whorls, counting from trunk base to tree tip.

As a tree ages, however, individual branches and even whole branch systems can be knocked off or shaded out. A more reliable way to estimate the age of temperate-zone trees such as oaks and pines is to count internal tree rings. What causes these annual circles? During the spring and summer, when sunlight and water are readily available, trees grow actively from the cambium, the living layer of cells just behind the bark. Individual cells manufactured during the warmer months are generally large and robust. In the fall, when the days shorten and water in the soil becomes less available due to freezing, tree growth slows and the size of new cells decreases. In winter, aboveground tree growth stops entirely. Through these cold months, trees maintain themselves on their reserves of sugars, which were generated during the summer by their leaves and then stored in their roots in the fall. The contrast between the earlier large cells and the later smaller cells establishes a ring that is visible in the cross-section of a tree trunk. The width of each ring expresses the tree's rate of growth during that year.

Dendrochronologists—people who study trees (*dendron*) and time (*chronos*)—have developed several ways to measure tree rings. One method involves extracting core samples from living tree trunks. A hollow tube of cold-cast steel with a threaded point at one end is inserted through the side of a tree, and a slender core sample (one-sixteenth inch in diameter) is extracted. Putting it under a dissecting microscope, the forest biologist counts the number of rings in the core, from center to bark, recording the width of each ring in the process. However, tree-ring dating is not always feasible or accurate, as when tree cores have rotted because of disease or when seasonal signals are weak or irregular, such as in the dry subtropical woodlands of southern Texas, where a long drought can shut down oaks during their typical winter growing season. Moreover, foresters are of mixed opinions about the possible harm inflicted by coring. Some argue that penetrating the cambium makes the tree vulnerable to viruses and bacteria, though others insist that a healthy tree can endure such an injury and recover the way a human recovers from a small puncture wound.

Measuring the age of trees in tropical forests has been a perennial problem for ecologists. Because in many places favorable growth conditions occur year-round, many trees in the tropics are constantly growing and so have no rings at all. Even trees that do undergo seasonal growth, such as those in habitats where rainfall is concentrated into a few months of the year, have unreliable rings because they can jump into a growth mode in response to even small inputs of out-of-season rainfall. The tropical dendrochronologist, therefore, must turn to other methods to determine a tree's age.

One surprisingly useful tool for tree dating emerged through the development of atomic weaponry. During the early era of nuclear testing, atomic devices were detonated in the atmosphere. The radioisotopes ejected from these explosions spread worldwide, forming a thin, weakly radioactive blanket over the earth. Some of these radioisotopes mimic naturally occurring elements so closely that many plants and animals cannot distinguish them. Trees take them up and incorporate them into their cells, along with their regular nutrients. Radioactive strontium, for example, mimics calcium, a nutrient that plants use to build new cell walls, much as animals use calcium to build bones. In 1954, trace amounts of radioactive strontium generated from bomb tests wafted through the air, dissolved in rain, entered the water cycle, were absorbed by roots, and then were incorporated into the living tissues of trees. This resulted in a short-lived but distinctive radioactive signal that has been held in the tissues of all of the trees living in the world at that time. Now, half a century later, scientists extract pieces of wood from tropical trees and note where in the cross-section of the trunk the "1954 bookmark" of radiation occurred. This allows scientists to measure how much each tree has grown since that time. Although the results cannot be extrapolated to determine how old a tree is, they do provide the dendrochronologist with an exciting tool to compare the rates of growth (from 1954 to the present) of individual trees and different species of trees that lack reliable rings. By revealing relative growth rates, this approach gives scientists a better understanding of population dynamics within forests.

Our knowledge of tree rings tells us more than simply how old a tree is. Tree rings can be used to explain a musical mystery as well. One day my daughter, Erika, who plays the flute in her middle school band, came home with a question from her teacher: Why are violins crafted by Antonio Stradivari in the seventeenth century so far superior to others created not just since then, but also in his own time? Every violin has a distinctive "voice" that is its own, and Strads are recognized as having the very best in the universe of stringed instruments. Musicians—and more recently, scientists—have sought to explain the superb sound quality of this luthier's products. They know that spruce wood was used for the harmonic top, willow for the internal parts, and maple for the back, strip, and neck, and that the wood was treated with several types of minerals in the form of varnish. Recently, Henri Grissino-Mayer, a dendrochronologist at the University of Tennessee, and Lloyd Burckle, a climatologist at Columbia University, suggested that the wood Stradivari used for his instruments developed special acoustic properties during the "mini ice age" that, peaking between 1645 and 1715, bestowed abnormally cold weather on much of Europe between about 1400 and 1800. This long period of extended winters and cool summers caused the slowest tree growth in the last five hundred years and led to the unusually dense, and resonant, alpine spruce that Stradivari used for his instruments. Why he achieved such astonishing brilliance of tone in his instruments, and other violin makers did not, remains a mystery, however.

EARTH'S OLDEST TREES

Thomas Browne wrote in 1658: "Generations pass while some trees stand, and old families last not three oaks." However, even he would have been flabbergasted to learn the age of the oldest trees on earth. On the windswept flanks of the White Mountains of California, bristlecone pines have hung on to life for not just tens or even hundreds of years, but millennia. Many of these trees were seedlings when the pyramids of Egypt were erected, and they gained maturity at the time of Christ. Their average age is a thousand years, but a few individuals are over four thousand years old. The oldest known bristlecone pine, appropriately named Methuselah, was determined in 1957 to be 4,723 years old, through a core sample. It stands today in silent company with its ancient siblings, which bear similarly reverential names such as Buddha and Socrates.

The difficulty of aging trees is nowhere more clearly demonstrated than in the story of Prometheus, one of the ancient denizens of the groves that we now know—posthumously—lived for over 4,900 years. Donald Rusk Currey, now a professor in the Geography Department at the University of Utah, carried out his Ph.D. research in eastern Nevada in the 1960s. He was seeking evidence—including the ages of trees—that could explain patterns of past glacier movement on the slopes of Great Basin National Park, and part of his work took place in a bristlecone pine grove at 10,000 feet. He chose the oldest-looking individual and drilled into it with the longest increment borer he could get (twenty-eight inches long), but he failed to get a clear reading and broke two of his corers in the process. He then approached the district ranger, asking—and being granted—permission to cut the tree down to examine the whole trunk in cross-section to determine its age. Currey learned that it had existed in that spot nearly five millennia; he also realized that he had cut its life short prematurely. Although today his action may seem surprising—if not sacrilegious or even downright criminal—at the time the experts had stated that the oldest bristlecone pines grew in California, not Nevada, and there was no external indication that this tree would be surpassingly older than its neighbors. Currently, the site where Prometheus was felled is unmarked, just a stump with some debris scattered about; no sign documents either Prometheus or the human desire to explore the hidden secrets of time and trees.

Just as the proverbial goldfish grows to fit its bowl, the stature of a bristlecone pine reflects its environment. Living in California in the windswept White Mountains, whose peaks rise above 14,000 feet, these trees are not large; indeed, they grow to a maximum of sixty feet in height and forty inches in diameter, and increase in girth only one-hundredth of an inch per year. Why is growth so slow? This subalpine zone can experience frost at any time of the year, and its growing season lasts only six weeks, typically falling in July and August when temperatures range between 40°F and 65°F. During this narrow slice of time, bristlecone pines must store reserves for the other ten and a half months of cold weather, and they thrive with only ten inches of annual rainfall. Over many generations, the tree has evolved unique ways to endure such conditions. In contrast to the needles of their relatively short-lived conifer cousins such as spruce and hemlock, which retain their foliage for three to five years, the needles of bristlecones can persist for up to thirty years. Longer-lived needles provide a stable energy-capturing apparatus that sustains trees during years of severe stress and reduces the amount of new materials the trees need to gather, much as a thrifty car owner will hang on to his old Ford to avoid having to write a large check to pay for a new model. When lightning or drought damages these trees, their bark and the injured part of the xylem—the tissue that transports water—die back. This survival mechanism reduces the amount of living tissue that the crown must maintain, just as our bodies conserve heat in very cold conditions by reducing the blood flow to our extremities to keep core organs warm. The surviving parts of the tree, fed by the narrow line of residual living xylem, remain healthy. One bristlecone named Pine Alpha, for example, nearly four feet in circumference, has only a ten-inch strip of living bark girdling its trunk, but it has lived more than four thousand years.

Bristlecone pines are not the only trees whose relatively diminutive size belies their advanced age. In the old-growth forests of the Pacific Northwest, some of the small trees that occupy the forest understory—many no more than twelve feet in height and five inches in diameter—are hundreds of years old. These so-called shade-tolerant trees, such as western hemlock, are able to germinate in the deep shade of the multi-layered canopies above them. In their youth and into middle age, they seldom grow taller than twice my own height, drawing on the small amount of energy that penetrates the canopy, only about 2 percent of the total sunlight. They can live like this for decades or even centuries until one of the huge overtopping trees dies from disease or is knocked down by wind, which creates a gap for light to reach the forest floor. Within a few years, the understory tree shoots up rapidly in response to the pulse of sunlight, at last occupying an upper-tier spot in the sunlight. It's like a teenaged boy, who can shoot up three shoe sizes in a single year but whose growth then stabilizes as he enters adulthood.

SEASONAL SHIFTS

> O Wild West Wind, thou breath of Autumn's being—
> Thou from whose unseen presence the leaves dead
> Are driven, like ghosts from an enchanter fleeing,
> Yellow, and black, and pale, and hectic red.
>
> —*Percy Bysshe Shelley, "Ode to the West Wind"*

The poet Shelley imagined that leaves changed their color as the wild west wind drove them "like ghosts from an enchanter." I sometimes think of wind and tree as instrument and dancer, her limbs arching and springing back, changing in curve and color from season to season.

The processes involved with leaf color change and leaf fall are indeed as complex and intricate as any ballet performance. Trees take up water from the ground through their roots and take in carbon dioxide from the air through their stomata, tiny holes in their leaves. Sunlight energy turns the water and carbon dioxide into sugar that can be either used immediately for growth or stored for times of cold and darkness. This process of photosynthesis, which means "putting together with light," requires the efficient absorption of sunlight. Most leaves produce several pigments that can absorb light, the best known and most abundant of which is chlorophyll, a complex molecule that is green in color. Other absorptive pigments are the carotenoids, which are yellow and orange, and the anthocyanins, which are red and purple.

As summer ends, the days grow shorter. Contrary to popular opinion, it is chiefly shrinking daylight, rather than cooler temperatures, that triggers the leaves to produce the hormones that, flowing from one part of the tree to another, function as internal traffic signals to regulate the flow of sugar, water, and nutrients. As the autumn equinox nears, the tree shifts from energy production and nutrient transport to energy and nutrient storage. The rate of daylight shift, based on the tilt of the planet, is a far more reliable indicator of time than temperature, which can vary greatly from year to year and region to region.

The dropping of leaves constitutes a large loss to the tree in terms of the materials and energy that went into making them. Yet trees—like people—do recycle materials. One mechanism by which trees conserve energy and nutrients is to disassemble the molecules they cannot use in the winter, transport them through their phloem—the innermost living tissue of the bark, which carries sugars and starches for energy to where it is needed—and store them away in their roots so that they can be used again the following year, when water and sunlight become abundant and growth can begin anew. During the shortening days of autumn, for example, the green chlorophyll molecules are taken apart and disappear from the leaves. This recycling of nutrients by trees is especially common in places where nutrients are limited by poor soils. As the bright green fades away, the vibrant yellow, orange, and red colors of the preexisting carotenoids and anthocyanins are revealed. Nutrient recycling is also called into play by trees living in soils that are extremely poor in nutrients, or when nutrients are present but locked in and unavailable for plant uptake, such as in pine barrens and peat bogs.

Seasonal tree transformations have major economic implications. For example, more than 1.5 million tourists visit Vermont each autumn to view the foliage, and they spend more than $700 million. In Wisconsin, a larger state, the fall colors bring in more than $1 billion. Those responsible for economic development in these states note that if the timing of the displays could be reliably predicted, even more visitors would come, with an accompanying boost in revenue. But forest biologists do not yet understand all the mysteries of tree dynamics, and so the guessing game continues.

MOVEMENT THROUGH TIME

Trees are dynamic at the level not only of individual leaves, but also of twigs and branches. Several years ago, I made a visual record of the movement of individual branches of a single tree, which allowed me also to quantify the collective movement of the tree as a whole. Tying a small paint-dipped paintbrush onto a twig, I then held up a piece of paper and allowed it to have contact with the dancing tip of the brush for two minutes. The image created by the tree's motions was highly aesthetic, reminiscent of Chinese calligraphy. I worked with a variety of trees and found that each species had a rather different "signature," based on the resilience of the twigs and the flexibility of the branch wood.

To quantify the movement of all of the branches, I measured the length of each of the lines painted by the single twig and added them up to get a collective line length for that twig for the two-minute period. I then counted the number of twiglets per branchlet and the number of branchlets per branch, and estimated the number of branches on the tree to arrive at a total number of twiglets. Finally, I multiplied the collective line length for the single twiglet by the total estimated number of twiglets, then by the number of minutes in a year, divided the result by two, and arrived at a total sum of twiglet movement per year. A hundred-foot tall Douglas-fir, for example, moves the astonishing equivalent of 186,400 miles in a year, or over seven times the circumference of the earth. These estimates not only give me a "gee whiz" factoid to put on parade but also provide insights into scientific questions relating to tree physiology and moisture balance of soils. The rate of movement of air greatly affects the rates and amounts of water taken up by roots and thus removed from the soil. Greater distances of leaf movement will result in more water sucked from soils; our "tree as artist" activity might well have a place in the toolbox of physiologists in the future.

Although we can describe the mechanisms of color change in foliage, and appreciate that branches may sway collectively for miles and miles, we think of trees as immobile. There are some striking exceptions to that rule, however, such as the stilt palm of lowland tropical forests. The upper trunks and crowns of these trees rise high above the forest floor, supported by a massive cone of stiltlike roots that develop from the trunk one or two yards from the ground and then arch outward to sink firmly in the soil, buttressing the trunk. On the dark floor of the rainforest, any method of growth that allows a tree to gain access to light will favor that tree and its offspring. In the case of these palms, roots grow not only downward but also out toward the light. The cone of stilt roots thus becomes asymmetrical, and over the course of a few years a fast-growing stilt plant can literally pick itself up and "walk" out from under fallen limbs, inching its way from a dark patch of the understory into a nearby light gap. The lower trunk and older roots rot away and are left behind as the tree moves onward. One mature individual, about fifty feet tall, grew a new root at a rate of nine inches per month until it reached the ground. Other roots grew horizontally, which led the tree from dark to light—a distance of 4 to 80 inches over a period of three years of observation. Alpine krummholz trees, too, are in a sense mobile, creeping laterally across the ground rather than battling the rugged constraints of snow and strong winds to try and grow upward.

In a forest stand, all this dynamism adds up. One summer, the poet John Calderazzo visited my old-growth study sites and was taken not so much by the size and girth of the five-hundred-year-old living giants, but rather by their fallen counterparts, laid out like Grecian columns on the forest floor, some entire, some snapped and buckled. For three days running he sat in one of the tree falls listening for their stories, which he eventually captured in a poem called "Douglas Fir, Falling."

Surely,
Somebody must hear one now & then,
A big tree falling on its own.
So why not me,
Hiking in the submarine green
Along Panther Creek among Douglas firs,
Their trunks as wide as my outflung arms,
Swaying wind in the rivering crowns
Almost drowning the steady
Breeze of creek water . . .

I just hear, behind the tangled wall
Across Panther Creek,
The building fury

Of the tree's descent,
Leaf & branch storms set loose
In the bird-panicked, lichen-torn air.
Swirling trunk dust, slammed
With earth, explodes
From the forest.
The ground quakes, the tree bounces
Once, cracks in three places.

Then everything seems to
Stop—creek water, canopy wind,
Rasping drizzle of needle litter
& shredded bark,
Even my own breath:
All of it on hold
As if to honor the tall life
Of this forest king,
Which has temporarily fallen back
Into the grand jumble of things.

BAROMETERS OF CHANGE

In the early days of tropical vegetation studies, explorers would journey into the rainforest with surveying tools, select a portion of the forest that appeared to be representative of the whole, and fell a line of trees. After taking precise measurements of the height to the tops of trees and to the base of the crown, of trunk diameters and lengths of branches, they would prepare a two-dimensional line drawing of thirty trees or so. The resulting "profile diagram" of that forest could then be compared to other such profiles. All were neatly tucked away for later viewing and analysis. Over time, wind, disease, fire, and other agents of disturbance wrought changes to the forest—but not to the diagrams. While the forest would show wear and tear, the paper profile diagram, unlike Oscar Wilde's portrait of Dorian Gray, remained pristine, reinforcing the image then held by most foresters of the forest as a place of stability.

In the 1960s, however, forest ecologists began to pursue longer-term studies of the tropics, returning to the same research plots year after year to gather comparative data. They marked individual trees with numbered tags and for thirty years re-counted the trees, keeping track of which ones toppled and by what agent. The forest proved capable of dramatic change over relatively brief intervals of time. Although individual trees might live more than three hundred years, turnover time—the average rate at which a given patch of forest opened up and was recolonized—was only about sixty years, a fraction of what the old explorers had surmised.

Similarly, tree-coring studies show that the long-lived ancient forests of the Pacific Northwest coast have been subject to multiple regimes of disturbance, including windstorms, insects, and floods, but primarily fire. The fire scars visible in the rings of some tree cores indicate that fires can sweep through areas and leave mature, well-established trees standing, just as an earthquake can leave a city in rubble, save for the few buildings constructed to withstand the wrenching of shifted foundations. Like those quakeproof buildings, the trees that can survive fires share certain qualities. One pattern includes trees that have evolved a "resister" strategy, relying on thick bark and a tall crown to survive moderate fires. Douglas-fir, the dominant species in these coastal old-growth forests, and ponderosa pine, an inhabitant of forests on the dry, fire-prone east side of the Cascade Mountains, are examples. If a severe fire sweeps through, all is not lost. On the contrary, because the seeds of these trees require a large amount of sunlight to germinate and persist, a disturbance regime that includes serious conflagrations and large windstorms is a requirement if they are to start a new generation. In contrast, many of Douglas-fir's co-inhabitants, such as western hemlock, are top-killed by fires and must be replaced by new saplings that sprout from seed. From there, they must struggle their way up through the canopy, waiting in shade, sometimes for centuries, before a neighboring large tree topples to provide a patch of light.

Dendrochronologists gather information from tree cores about past events such as fires and periods of warming or cooling that can provide humans with long-term records of climate. Andrew Ellicott Douglass, the grand old man of dendrochronology, was the first to note that the wide rings of certain species of trees were produced during wet years and narrow rings during dry spells. He also discovered that he could match core samples taken from dead trees of unknown age with samples from living trees whose rings were dated, thereby expanding our knowledge of growth and climate beyond a single generation of trees. In 1937, Douglass established what is now the mecca for tree-ring scientists, the Laboratory of Tree-Ring Research at the University of Arizona. There, an international cadre of researchers sits hunched over microscopes, using tree rings to reconstruct past climate regimes worldwide.

In an intriguing aside, tree rings are oddly analogous to the ear bones of fish, or otoliths (from *oto-*, meaning ear, and *-lith*, stone): tiny crystalline plates that aid in the hearing and balance of fish. Otoliths grow concentrically from inside to outside, exactly like the rings of a tree, with light and dark bands representing periods of high and low growth on yearly, monthly, or even daily cycles. Because the width of the ring—indicating the amount of growth—has a direct relationship to water temperature, otoliths provide quantitative information on key environmental conditions. Otoliths of arctic char, landlocked fish that dwell in the far north of Canada, for example, are now used as indicators of global climate change. These tiny bones also contain, in minute concentrations, the chemical constituents of the surrounding water, traceable through time. In this way they provide a virtual "diary" of a fish's early life, allowing biologists to trace a fish's journey between different habitats—from a polluted nearshore estuary, say, to cleaner waters.

Palynology, the study of pollen records, provides scientists with another way to understand long-term dynamics of whole plant communities and their past environments. Pollen is well suited to this topic because each species of plant has a uniquely identifiable structure; moreover, the outer shell of pollen is composed of a highly resistant compound called exine,

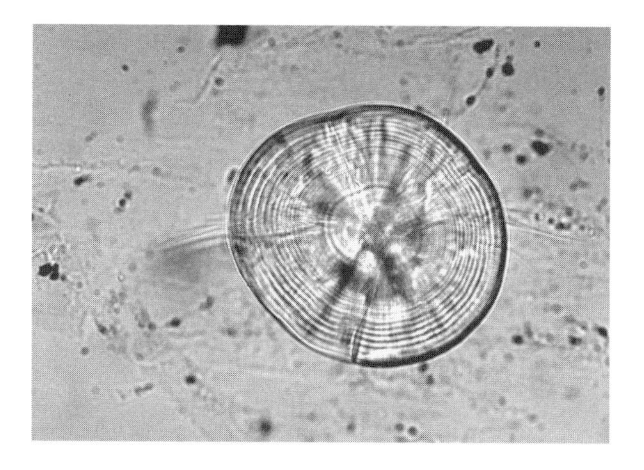

21. This otolith, a bony structure extracted from the ear of a fish, shares similarities with trees, in that the fish lays down annual rings just as a tree does. The relative width of the otolith rings can indicate past environmental conditions, just as the cross-section of a tree can. Otolith from a larval haddock reproduced by permission of Dr. Steven Campana, Bedford Institute of Oceanography, Canada.

which remains unchanged over centuries, unaffected by the forces of decomposition that reduce other organic materials to their constituent parts. The grains are very light (the term *pollen* comes from the Latin for "fine flour") and as trees release their pollen, the grains float through the air, settle on the surfaces of nearby lakes and ponds, sink to the bottom, and eventually become covered with sediments that build up over the centuries. Palynologists row out to the middle of these lakes, loaded down with the same metal coring devices that dendrochronologists use. Plunging the augers vertically into the lake bottom, they extract long tubes of layered sediment, which are then frozen and thinly sliced, allowing for examination under a microscope. Researchers (or more commonly, their graduate students) identify and count individual pollen grains to construct a picture of which species of plants grew in what proportions around the lake thousands of years ago.

In the Pacific Northwest, tree rings, pollen diagrams, and computer simulations of the earth's atmosphere indicate that tremendous fluctuations in temperature and rainfall have occurred over time. Thirty thousand years ago, vegetation west of the Cascade Range was made up of open forest tundra interspersed with lodgepole pine and Engelmann spruce. By 10,000 years ago, the continental ice sheet had collapsed and summer sunlight increased because of changes in Earth's orbit. Pollen records for that time show a predominance of Douglas-fir, oak, and prairie herbs, species that rely on hot weather and frequent fires to thrive. Indeed, lake sediments show a marked increase in charcoal concentrations as well. By 6,000 years ago, moister climatic conditions were established, mirrored by decreases in the pollen of Douglas-fir and grass and increases in the pollen of western red cedar and western hemlock—the two predominant trees species of the Puget Sound lowland today. Thus, the Douglas-fir stands we call "ancient forests" are, in a geological context, mere adolescents.

FOREST PROGRESSION AND SUCCESSION

In primary (uncut) forests, an area the size of a football field might contain as many as fifty dead standing trees, or snags. These snags may stand for half a century. However, death is not the end of the line for trees, which provide shelter for the living. A progression of decay occurs in dying, dead, and fallen trees, with a diverse array of animals, plants, and fungi taking advantage of the various stages. The first to inhabit declining trees are fungi, followed by bacteria, yeasts, mites, and nematodes. Then the food chain adds more links, including, in the Pacific Northwest, more than eighty species of invertebrates. Some insects move in when the dead tree is fresh and hard, others when it is soft and yielding.

Birds also use the successional sequence of snags, starting with the powerful pileated woodpecker. The holes that it drills in ailing but still standing trees shelter the bird itself, followed by secondary cavity nesters, who must use tree holes excavated by other species. The pileated woodpecker

22. *Dead logs on the forest floor were previously discounted as "wasted wood," but ecological research shows that they provide critical habitat for myriad plants and animals. Photo by author.*

is a keystone species, helping with its pioneering work numerous other birds, such as nuthatches, chickadees, bluebirds, and swifts, which roam the decomposing, riddled trunk like teenagers in the food court of a suburban mall. As more space opens, mammals arrive, including squirrels, fishers, martens, wood rats, and black bears. Most bats were assumed to live in caves, but recent studies show that many roost in abandoned woodpecker holes in aged trees. A single ponderosa pine tree in the Coconino National Forest of Arizona, for example, housed over nine hundred Arizona occult bats.

Eventually, the snag falls and becomes a member of the forest floor, now termed "coarse woody detritus," an academic's name for a dead log. At this point, a new phase begins. Rodents use these logs for shelter and protected travel. Owls have been seen lurking at an end waiting for an unsuspecting mouse meal to emerge. Many amphibians need the humid woody debris for feeding, reproduction, and a hiding place. Logs also support gigantic colonies of carpenter ants, which in turn supply nearly all of the food for the pileated woodpecker and are also a critical portion of the diet of the black bear.

And what about the aquatic world? Forest managers used to think that deadwood clogged streams and hindered the passage of fish. They spent large amounts of money removing logs from streams to clear the way for salmon to make their quadrennial journeys between headwaters and ocean and back again. In the past decade, however, aquatic biologists have learned that deadwood in streams is a critical environmental component and must be in place if fish are to spawn successfully. These logs, which can remain intact for over a thousand years, collect sediment and provide protection from floods so that plants can recolonize washed-out sections of the river. In-stream wood, along with the slime that covers its surface and the leaves and needles it traps around it, provides habitat for aquatic fungi, algae, and invertebrates, all of which provide food for fish. The small dams made by fallen trees account for up to 80 percent of the deep pools in streams, features that salmon need to rest and forage in on their way upstream to mate, lay eggs, and restart their life cycle.

When logs wash down to estuaries and tidelands, they provide critical perches for kingfishers, eagles, and other predators. Once wood hits salt water, it encounters another set of consumers—voracious marine invertebrates that eat wood. These include many species of shipworms, which are not actually worms but marine mollusks that tunnel through deadwood. Other animals benefit from the feasting shipworms as well. When large pieces of submerged driftwood roll along the bottom of the ocean during winter storms, they gather in submarine woodpiles. These eventually turn into biological hotspots, teeming with a marine menagerie of predatory limpets, small fish, and snails that use the powder and feces of the tunneling shipworms. Larger fish, such as carnivorous tuna, which feed at the top of the food chain, follow the smaller fish, which feed off even smaller fish and invertebrates that derive resources from these chunks of driftwood, which can stay afloat for up to five years and travel thousands of miles. A single piece can thus develop into a nomadic marine supermarket, with shoppers that include birds along with sharks and other fish. The dead log created on land constitutes a mini–Noah's Ark, made, like the original ark, from wood.

The value of fallen trees is far-reaching. Yet even as we learn about the tight relationships of life to deadwood, humans are endangering these functions by removing too many dead or soon-to-be dead trees for our own uses. With the increasing harvest of entire forests for lumber and pulp, the source of large snags and fallen logs is disappearing. In the United States, the old-growth forests are nearly gone. Meanwhile, salvage harvesting of stumps, logs, and driftwood—in the name of "forest health" and security of homes against fire—is denuding our landscape of habitat and sources of food for the wild creatures of the earth.

How that vacuum will be filled is a rising concern to those who wish to maintain a seaworthy vessel for all species of animals. Ever since the significance of deadwood became clear, modern-day Noahs—in the form of ecologically minded forest managers—have been creating surrogate deadwood in ways that might simulate the functions of the real thing. In the Southeast, where old long-leaf pine trees are now scarce, biologists are trying to jump-start the decomposition process by injecting fungi into young trees, thereby enhancing habitat for the later-colonizing species. Washington State now requires loggers to leave a certain number of snags and logs and to preserve a narrow uncut strip along rivers and streams. Where large logs are in short supply, forest managers are making bundles of logs of small diameter, hoping to mimic the benefits of old-growth deadwood. After one hundred years of pulling wood from rivers, the U.S. Army Corps of Engineers now anchors poles in Pacific Northwest streams, at great expense. No one can yet say whether these techniques will prove effective, but they do seem a step in the right direction—toward a more sustainable balance of life with death.

LESSONS FOR A NEW MILLENNIUM

Distant, fragile, pale blue of sky
Above the forest trees lifting
Their arms, a silent offering of green.
Winter's tender sun on earth
Spreads a shawl of clear light.
I am writing this down before
The painter, indifferent, wipes clean the canvas.

—Rabindranath Tagore, "Recovery 6"

Trees remind us of the need to accept change even if it seems destructive, and even if it raises the specter of our own mortality. Nearly three decades ago, I established a set of permanent research plots in the tropical cloud forests of Monteverde, Costa Rica. My students and I assiduously identified, measured, and tagged 2,489 trees, and every five years we return to recensus those plots. In between our visits, huge trees fall over, creating large gaps in the forest, open spaces that are then colonized by seedlings. Although these gaps evoke in us a sense of death and destruction, the light-filled expanses are necessary to enable the next generation of forest plants to flourish.

By planting trees, we can foster their renewal and at the same time extend our own imprints beyond our lifespans. In 1732, Thomas Fuller wrote: "He that plants trees loves others beside himself." This is true everywhere, of course, but in Israel the planting of trees has meant more than simply placing a sapling in the earth, with the promise of shade or fruit. There, tree planting was extolled as a way to "make the desert bloom." Now, that simple action is bound up in interesting ways with religion (planting a tree in the Holy Land) and nationalism (a means of staking claim to the land), and as a way for Jews abroad to demonstrate community with Israel. Because of the longevity of trees and the way they link the earth to heaven and people to the land, planting a tree signifies that the people will be in a place for a long time and that they foresee their children and their children's children dancing in its shade, climbing in its limbs, leaning quietly against its trunk.

The way we think about trees and time is changing, however. What we know now about the age of trees, how trees reflect environmental changes, and the complex relationships of trees with other organisms through time has forced us to think differently about human impacts on trees and forests than we have in the past. Ecological journals and, increasingly, newspaper front pages and television screens are filled with predictions of dire consequences from forest fires, deforestation, forest fragmentation, and global climate change.

The ways we obtain and use energy are altering the global climate at an unprecedented rate. The pollen diagrams that documented tremendous long-term climate changes in the Pacific Northwest over the past 30,000 years show that trees were able to keep pace with this change—migrating hundreds of miles across the landscape—because it proceeded gradually. In contrast, the human-caused global warming that is occurring today will likely lead to very serious effects, such as the disappearance of polar ice and coastal inundation due to sea level rise, within as little as two or three decades—a thousand times faster than in the past. Even the most rapidly growing tree with the most mobile of pollen and seeds or most active of stilt roots cannot keep up with that rate of change. Climate shifts are outrunning the trees.

As rain and snowfall patterns change and drought becomes increasingly common, forests in many areas are being hit hard. Water-stressed conifers, for example, exhibit reduced resistance to bark-boring beetles. Beetle attacks and higher temperatures lead to more frequent and more intense forest fires, far in excess of the natural fire regime.

Another factor in the health of forests worldwide is deforestation and development. In the past, naturally vegetated landscapes were contiguous, which allowed for free movement in all directions. Today's forests, in contrast, tend to be highly fragmented and isolated. Nearly all of our landscapes outside of national parks and wilderness refuges are checkerboards of suburban, agricultural, and industrial regions. Over time, this process of fragmentation takes a serious toll on healthy ecosystems. From the standpoint of the physical environment, fragmentation increases edge effects. A tree that stands at the edge of a forest is exposed to far more wind, greater humidity extremes, and more sunlight than a tree in the middle of the forest. Fragmentation can also isolate populations of organisms from their own kind. Many animals with poor flying ability will not cross an open clear-cut or pasture that separates two patches of forest. This can slow or completely stop the flow of genetic exchange that maintains healthy variability in populations.

Even as these changes occur, we are racing to learn as much as possible about the past in order to prepare for even greater changes in the future. How will forests and their denizens fare? It is impossible to say—but with sound information from science, let us hope that our awareness of time and of Hesse's "narrow years and luxurious years" will ultimately steer our actions up Maslow's modified pyramid to mindfulness.

Chapter Seven

SIGNS AND SYMBOLS

A poem containing a tree may not be about a tree.

—*Marvin Bell, "Thirty-two Statements about Writing Poetry"*

One summer evening last year, I was standing in line to buy movie tickets. The cinema was showing three films—a historical drama, a motorcycle racing film, and a teenage horror flick—and a diverse file of people stretched ahead of me. I had spent that afternoon reading papers from my "Trees and Humans" class about symbols that are associated with trees. Looking down the movie line, I wondered how my students' conclusions compared with what these people might think. So I drew a picture of an oak tree and walked down the line, explaining that I was a college professor interested in trees and asking them what they associated with the image of a tree. Their responses were remarkably similar to those of my students: beauty, life, the sacred, usefulness, familiarity, sturdiness, something to trust, wisdom, old age, life, death, strength, lumber, peacefulness, "myself." My little experiment supported what the literature on tree symbolism documents, that trees are powerful and broad-ranging symbols for just about everyone.

OURSELVES AND FAMILY

I wanted to see the self, so I looked at the mulberry.
It had no trouble accepting its limits,
yet defining and redefining a small area
so that any shape was possible, any movement.
It stayed put, but was part of all the air.
I wanted to learn to be there, and not there,
like the continually changing, slightly moving
mulberry, wild cherry and particularly the willow.
Like the willow, I tried to weep without tears.
Like the cherry tree, I tried to be sturdy and productive.
Like the mulberry, I tried to keep moving.

—Marvin Bell, "The Self and the Mulberry"

The most basic question that humans ask is, Who am I? The most interesting rejoinder of my movie line respondents was that the tree drawing symbolized the self. We both have limbs, trunk, and crown. Marvin

23. *This striking piece of art portrays the intimate ways that humans and trees are bound together. Louise Duffy,* Listening to the Tree, *reproduced by permission of the artist.*

Bell's poem expresses other ways we are like trees: we weep without tears, we are sturdy and productive, we move in obvious and subtle ways.

Many of us have trees in our names, which may tell us where our forebears came from or how they made their living. Many family names have links to trees, both generally—Wood (English), du Bois (French), Skov (Danish), and Holt (Dutch)—and specifically—Li (plum tree in Chinese) and Perry (pear tree in Old English). Surnames may be related to places associated with trees—Ashley (ash tree clearing, Old English), Greenberg (green mountain, German), Oakley (place of many oaks, English), and Ogden (acorn, *âc*, and valley, *denu*, Old English). Some surnames reflect tree-related occupations, including Carpenter, Sawyer, Woodward (ward or guardian of the wood), McIntyre (son of the carpenter, Scottish), and Zimmerman (carpenter, German).

Just as people took on the names of the trees that surrounded and nourished them, so they also named their communities after them. In the Old Testament, for example, many places are identified by their trees. In Joshua 19:32–34, we read: "Their boundary went from Heleph and the large tree in Zaanannim . . . and ending at the Jordan." Today, we see the same tendency. Pulling out a map of a random state, I am sure to find many places named for trees. In Idaho alone, the name of the capital city, Boise, derives from the French word for forest, *bois*. Other tree-related town names in that state are Yellow Pine, Pinehurst, Fernwood, Tamarack, Greenleaf, Fruitvale, Cottonwood, Woodland, Dubois, Myrtle, and Fruitland. Builders of retirement communities and housing developments often capitalize on our positive associations with trees, giving them such names as Shady Acres, Willow Estates, and Orchard Homes.

Trees have also been used to mark the bounds of home territories. When measuring off land parcels, settlers of forested lands carved the year in tree bark, and this permanent mark on what came to be called "witness trees" lasted across decades and landowners. My in-laws raise beef cattle on former pinewood land in southern Florida. When my family visits the ranch and walks the oak-palm hammocks that are scattered through the pastures, we look for relict witness trees, the grooves of date and identification number worn but still legible, part of the work of mapping what was then wild land, with not a cow or a swimming pool to be seen.

Even saplings or sticks, carefully placed, can deliver complex messages. Writing about the indigenous people of the forests of Borneo, the anthropologist and geographer Wade Davis, in his book *Shadows of the Sun*, describes "sign-sticks," "branches or saplings decorated with symbols that [maintain] an open dialogue between . . . widely separated groups" of the hunter-gatherer Penan people. The sign-sticks might tell

> where and when a party had split up, the direction each group had traveled, the difficulty of the terrain, and whether or not food was available. A large leaf at the top showed that the stick had been left by the headman. A folded leaf told that the group was hungry, in search of game. Knotted rattan gave the number of days anticipated in the journey. Two small pieces of wood placed transversely on the sign-stick indicated that there was something for all Penan to share, that all people of the forest were of one heart.

EVERYDAY LIFE

The moviegoers I queried associated trees with things that affect our daily lives, a frequent theme of literature. The pattern emerges in one of my favorite books, Italo Calvino's *The Baron in the Trees.* This story, set in eighteenth-century Italy, describes the decision by a rebellious young nobleman, the baron Cosimo, to protest a minor injustice committed by his father by spending the rest of his life in the treetops. After a few months of adjustment, his arboreal existence becomes completely natural, not just to him, but to his earthbound family and the local townspeople as well. He maintains relationships with his lover, Viola, and his dachshund, Ottimo Massimo. He even reads novels to an imprisoned brigand, illiterate but literarily disposed, as he sits perched in a tree neighboring the prison tower. He establishes a system of arboreal water pipes to help village firemen, invents a technique for treetop fishing, and locates numerous romantic spots for canopy-level trysts with Viola. In a delightfully matter-of-fact way, Cosimo makes the participation of trees in the daily rounds of people apparent and natural. Although modern humans do not depend on trees quite so thoroughly as the baron does, we certainly surround ourselves with trees—and symbols of trees—in our everyday life.

They are, for example, a rich source of figurative language. When we prepare for a movie date, it's with the "apple of our eye," and we get "all spruced up." When we settle down, we "put down roots," and when we need to start over, we go "out on a limb," "branching out" into new endeavors, perhaps even "turning over a new leaf." When we seek answers to "knotty" problems, we may find ourselves "barking up the wrong tree," and if we can't figure out a solution, we feel "stumped." When we are fearful, we "shake like a leaf." When we go home at night, we "sleep like a log." In myriad ways, we take on the attributes of trees.

Conversely, we give the trees our own attributes, personifying them, usually as humanlike creatures who live in and protect trees. In the classical world, hamadryads were female spirits who lived in enchanted groves, guarding the trees, and who died if the trees were cut down. Often depicted as wisps of light, dryads are playful creatures who may help, hinder, or tease humans. In Japanese folklore, where all phenomena of nature manifest different divinities, Kuko-no-chi is a deity who lives in the trunks of trees, and Hamori is the protector of leaves. Often in Japan, in the hollow of the tree, a tiny chapel is established where the faithful leave offerings. The beautiful Swedish wood nymph Skogsrå is the guardian of the woods and wild animals. In the biblical Song of Songs, a beloved woman is described to be "in stature like the palm tree, its fruit clusters [her] breasts" (7:8–9). She says, "As the apple tree among the trees of the wood, so is my beloved among the sons. I sat down under his shadow with great delight and his fruit was sweet to my taste" (2:3). In Greek myth, the mortal maiden Daphne begged the gods for protection from the overly amorous Apollo and was transformed into a laurel tree.

Personification is connected to the age-old tradition of "knocking on wood" to ward off bad consequences when commenting on good fortune ("We haven't had any problems with the car so far, knock wood"). Some linguists believe that this custom came about to invoke the aid of the spirits that dwelt in sacred trees such as the oak or ash, while others point to an old Norse belief that by knocking loudly on a tree while making a bold statement, the speaker could prevent the spirit of the tree from hearing, thereby keeping it from interfering in the speaker's plans.

Some tree symbols have more recent origins. For example, tying a yellow ribbon around a tree as a sign of loyalty to family, friends, or loved ones who have been away a long time, especially under difficult circumstances such as war or prison, was inspired by a 1973 hit song, "Tie a Yellow Ribbon 'Round the Ole Oak Tree." In the song, the ribbon was a signal to a man returning home from prison that he was still welcome; today, it is used to raise public awareness and reduce the feeling of helplessness of those who remain at home waiting.

And some traditions mix the old and the new. When traditional Scandinavians built a house, they placed a small conifer tree on the uppermost crossbeam as homage to their ancestors, whose spirits were said to live in the trees that now served as shelter. This assured them that the displaced spirits would continue to have a home. That tradition persists among carpenters and ironworkers today. For example, in 2004, the Philadelphia Phillies conducted a "topping out" ceremony at their new ballpark, the Citizens Bank Park, to celebrate the near-completion of the stadium's steel frame. The evergreen tree affixed by the ironworkers symbolized the hope of growth and prosperity for the new structure, and by extension for the team.

CELEBRATIONS

Tree symbolism emerges in human celebrations as well, gracing the calendar in every season. In spring, when plants bud out, May Day is celebrated. Its symbol is the maypole, traditionally made of hawthorn or birch, which is festooned with flowers and greenery. Today, schoolchildren perform dances around it, weaving ribbons in and out to create patterns. Originally, however, the maypole, with its phallic overtones, was probably a fertility symbol associated with the ancient pagan festival of Beltane, which is linked with the moving of cattle back into the fields from their winter housings and the rebirth of the land and living things. The blooming of the first hawthorn was taken to indicate the arrival of spring, setting the date of the Beltane celebration. The maypole traditionally appears in the United Kingdom, Austria, Hungary, Sweden, and Finland. Some associate it with Yggdrasil, the "world tree" of Norse mythology, a gigantic ash tree and the symbolic *axis mundi* that links the underworld and the heavens.

As the light green foliage shifts to the full deep green of summer tree crowns, other tree-associated holidays take their place in the calendar of human celebrations. Summer holidays that directly or indirectly involve trees occur in many cultures of the northern temperate zone, all with the similar theme of celebrating the longest day of the year; several of these include the burning of dead trees in bonfires. In Europe, many rites of this holiday are connected with nature and fertility, and young men and women jump over the bonfire flames. In Wales, May 1 is a holiday called Calan Mai or Calan Haf, the first day of summer. Celebrations start with bonfires in the public square. Golowan is the Cornish name for the midsummer celebrations in Cornwall. Historically, the celebrations were conducted from the 23rd to the 28th of June each year. The celebrations were centered around the lighting of bonfires and fireworks and the performance of associated rituals. Since 1990, the contemporary Golowan festival has revived many of these ancient customs and has grown to become a major arts and culture festival in late June. Other European countries that celebrate similarly in modern times are Ireland (where it is called St. John's Eve) and Sweden (where it is called Walpurgis, Valborgsmässoafton, or Valborg).

Analogous holidays exist in Slavic countries. July 7, Ivan Kupala Day, is the day the summer solstice is celebrated in Russia and the Ukraine. This is a pagan fertility rite that has been accepted into the Orthodox

Christian calendar, in which it has been associated with John the Baptist. Jāni is a Latvian festival held on the summer solstice, and people usually spend the day in the countryside. Jāni is thought to be the time when the forces of nature are at their most powerful, and the boundaries between the physical and spiritual worlds are thinnest. In the past, evil witches were believed to be riding around, so people decorated their houses and lands with branches and thorns of rowan trees to protect themselves. In modern days other decorations are more popular, including birch or sometimes oak branches and flowers. Women wear wreaths made from flowers, and men wear oak leaves; in rural areas livestock are also decorated. Another important detail is the festival fire, which must be kept burning from sunset till sunrise.

Autumn brings tree-related holidays as well. In fall, an important Jewish holiday, Sukkot (translated as the Festival of Booths), commemorates the Israelites' wanderings in the desert following their exodus from Egypt, during which time they lived in portable shelters or booths. In addition to marking a central event in Jewish history, this seven-day holiday, which starts on the fifteenth day of the lunar month of Tishri (late September to late October), celebrates the end of the autumn fruit harvest. Those observing the holiday construct, eat in, and even sleep in a *sukkah*, a temporary structure covered with a roof of tree branches. They also lift and shake branches of four species of trees: palm, myrtle, willow, and citron or etrog (a fragrant citrus fruit). Both the commandment to dwell in the *sukkah* and the taking of these four species come from the book of Leviticus (23:40, 42): "You shall dwell in booths seven days, all citizens of Israel shall dwell in booths. . . . You shall take on the first day fruit from a hadar [etrog] tree, branches of palm trees, boughs of leafy trees, and willows of the brook, and you shall rejoice before the Lord seven days."

In the dead of winter, evergreen trees become especially potent symbols. Two millennia ago, the Romans brought evergreen branches into their homes for the winter festival of Saturnalia, celebrating the renewal of life at a time when days are shortest and coldest. In A.D. 600, Pope Gregory I told the Christian clergy to encourage folk customs where Christian interpretations could be made, such as the use of greenery. The modern custom of the Christmas tree has been traced to sixteenth-century Germany, where fir trees decorated with sweetmeats were erected in village guild-houses on Christmas Day. Since then, mistletoe and holly have joined evergreen trees as emblems of the yuletide holidays. Many cities put up Christmas trees for everyone to enjoy. I grew up in Washington, D.C., and the most cherished end-of-the-year treat for my siblings and me was the lighting of the National Christmas Tree, just south of the White House. I gazed in wonder at the lights and limbs stretching up to the dark winter sky, feeling linked, through the tree, to both earth and sky.

Although most Christians have a Christmas tree in their homes on December 25, some pastors have been reluctant to erect Christmas trees in the church sanctuary because of the symbol's pagan origins. Instead, the Jesse tree, described in Isaiah 11:1—"A shoot will spring forth from the stump of Jesse, and a branch out of his roots"—is used, symbolizing Jesus' family tree (Jesse was King David's father). Jesse trees are depicted on banners and posters, with symbols of incidents in the life of Jesus hanging from the branches. Another Christmas tradition is the yule log, the central trunk of a large tree, which is placed in a pit and burned, bringing light and heat to those who gather around it, hands outstretched to the warmth.

In Costa Rica, people use canopy-dwelling plants to celebrate Christmas. Booths in the Mercado Central, the outdoor market at the center of San José, display large, fragrant bales of moss that have been collected from the cloud forests of the mountains. Mothers and grandmothers come by to purchase half a kilo of the green material to spread on the floor of the crèche that they make for their families. They then reverently place the donkey, the cow, the sheep, and the holy family on the mosses that until a few days before clung to the stems of windswept forests.

In Japan, people mark the New Year week by putting two pine trees of the same height in containers on each side of the front door. According to Shinto tradition, divinities live in the branches of trees, so this is a way to attract them and their blessings to one's home.

In the United States, Arbor Day is a holiday that specifically celebrates trees. Somewhat incongruously, it has its roots in Nebraska, which in the 1800s was a treeless plain. A pioneer couple, J. Sterling and Caroline Morton, arrived in that state in 1854 and brought with them a love for trees, which they planted around farms in the area as windbreaks and to provide shade, fuel, and soil protection. J. Sterling Morton wrote: "The cultivation of trees is the cultivation of the good, the beautiful, and the ennobling in man." After many years of advocating for trees, Morton proposed an official tree-planting holiday, which became Arbor Day and is now celebrated on the last Friday in April. Today, many states and other countries acknowledge a special day or week to plant trees and recognize the important roles that trees play in our lives. In his 1907 Arbor Day message, Theodore Roosevelt said:

> It is well that you should celebrate your Arbor Day thoughtfully, for within your lifetime the nation's need of trees will become serious. We of an older generation can get along with what we have, though with growing hardship; but in your full manhood and womanhood you will want what nature once so bountifully supplied and man so thoughtlessly destroyed; and because of that want you will reproach us, not for what we have used, but for what we have wasted.

In 2001, I lobbied Washington State to create Forest Canopy Week. We observe International Pickle Week in May, National Animal Shelter Appreciation Week in November, and National Park Week in April, so why not dedicate a week to the forest canopy? I wrote up a proclamation for Governor Gary Locke to sign, proposing July 17–24 for this special week, dates that encompassed the first moonwalk and the birthday of Sir Edmund Hillary, who first climbed Mt. Everest. This linked the canopy with other explorations of unknown places. The signed proclamation, with impressive curlicues and flourishes, now hangs on the wall of our Canopy Laboratory, and it is displayed proudly each year when we stage presentations and tree-climbing demonstrations to commemorate this special week.

EMBLEMS OF THE GROUP AND NATION

Trees figure large in the symbols we choose to represent our collective groups—families, tribes, companies, states, and countries—in the form of heraldry, business logos, postage stamps, currency, and flags.

Heraldry and coats of arms identifying individuals, clans, and lineages evolved in feudal Europe and gained popularity during the Crusades as a means of distinguishing warriors on the battlefield. Flowers, fruits, and leaves of trees are common heraldic motifs, with a variety of meanings. The inclusion of oak leaves and acorns, for example, signifies the strength and steadfastness of the bearer of the shield; the palm tree connotes righteousness and resurrection; the olive twig is shorthand for peace and concordance. Nations, too, use emblems such as coats of arms as icons of pride and representations of national culture, and often trees play a special role. The coat of arms of Barbados, for example, features the bearded fig, a tree that was common at the time of European settlement, while Italy's national symbol links the olive and oak, indicating strength and integrity in both peace and war.

Living trees are used to symbolize countries as well, in the form of "national trees." In 2005, China bestowed this honor on the gingko, a "living fossil" long used in traditional medicine. Costa Rica's national tree is the guanacaste, symbolizing stability and growth. The national tree of the United States is the oak, selected in 2004 in an online election hosted by the National Arbor Day Foundation. A press release explained the choice:

> Advocates of the oak praised its diversity, with more than 60 species growing in the United States, making oaks America's most widespread hardwoods. Throughout America's history, oaks have been prized for their shade, beauty,

> and lumber. They have also been a part of many important events, from Abraham Lincoln's use of the Salt River Ford Oak as a marker in crossing a river near Homer, Illinois, to Andrew Jackson taking shelter under Louisiana's Sunnybrook Oaks on his way to the Battle of New Orleans. In the annals of military history, "Old Ironsides," the USS *Constitution*, took its nickname from the strength of its live oak hull, famous for repelling British cannonballs.

Each U.S. state has chosen a species of tree to represent itself as well. Some of these trees evoke a very strong sense of place—the magnolia of Mississippi, the bald cypress of Lousiana, the Sabal palmetto of Florida, and Colorado's blue spruce. Pine is the most common genus of state tree, followed by oak, maple, and poplar. My home state of Washington, which is called the Evergreen State, invited the western hemlock to represent it in 1892.

The use of tree symbolism in the business world is ubiquitous as well—but to what extent I didn't realize until a few years ago when a student of mine, Jade Leone Blackwater, decided to collect tree-related logos and analyze their meaning. Within a few days of searching magazines, the Web, and billboards around town she had found over two hundred such logos, which followed a few general trends. First, trees often symbolize growth—physical, spiritual, or economic—that in turn is associated with protection, healing, and prosperity. Logos using trees in this manner tend to represent childcare and educational programs, medical practices, religious groups, and investment funds. For example, Sherwood/Bowman, Inc. is a company active in the insurance and investment industry that wishes to associate itself with long life, stability, and growth. Its logo is a handsome and healthy tree, rooted firmly in the ground, with flourishing bright green foliage. Second, trees are used to provide a sense of place, sometimes suggesting a particular city, in the case of a local business, or conveying a more general notion of "home," for companies that supply products used in and around the home. The Beirut outlet of the Hard Rock Cafe, for example, a company with restaurants all over the world, uses a cedar of Lebanon in its advertising imagery, which lends an identity and "locality" to this multinational business. Third, trees may be used to express concern for the environment or sustainability, especially in this time of growing consumer awareness of environmental problems. One example is DoubleTree, the major hotel chain, which is represented by two trees with intertwined crowns. In addition, its website prominently displays a link with the National Arbor Day Foundation, thus reinforcing DoubleTree's desire to appear committed to sound environmental policies and practices. Finally, companies that produce products associated with the outdoors, such as lumber, outdoor gear, and ecotourism, are often branded with trees. The U.S. Forest Service, which manages our nation's forested lands, sandwiches a conifer tree between the USFS initials for its logo. At a local level, the Michigan Conservation Districts use three conifer trees against a hilly backdrop to represent their work.

These trends apply not only to specific logos, but also to the way that advertising and media people promote their products. I learned this through one of the ongoing projects in my Canopy Lab, the TAI, which stands for the Trees Ad Investigation. Students skim through stacks of popular magazines that we garner as discards from dentist offices and the public library. We note which advertisements contain images of trees and forests, the product and company, and how the advertisement uses trees on the page. From this media analysis, we have concluded that companies use trees to sell a wide range of products that have little to do with nature itself—from SUVs, shown bashing through forested streamsides, to cigarettes for smokers who lounge casually in the shade of a sycamore. It is the qualities of trees that we treasure—strength, health, longevity, relaxation, and beauty—which advertisers invoke to sell their goods and services.

Another symbolic connection between trees and economics involves the designs on currency, both coins and bills. When the use of coins began over two millennia ago, their markings typically identified the current ruler or divine authority. Coins from the city-state of Athens in the early fifth century B.C., for example, depicted the patron goddess Athena on one side and her attribute the owl, symbolizing wisdom, on the reverse. Today, however, international coinage presents us with a wide array of culturally significant imagery—including trees, in the case of over forty countries, from Cyprus to Sri Lanka, Lebanon to Lithuania. Many of these depictions are simple and symmetrical branches or tree forms that encircle the text or figure on the coin. Some tree-imbued coins, however, tell a story. For example, the 1995 Lithuanian 50-litu coin shows not a thriving tree, but a stump, out of which, however, a strong branch is growing in full leaf. Issued on the fifth anniversary of the reestablishment of the Republic of Lithuania, following fifty years of rule by Russia, the coin symbolizes regrowth of a national identity after decades of repression.

Paper money originated much later than coins, in A.D. 810 in China, where pulp-based paper, printing, and ink were all invented. Most banknotes now are printed on heavy paper made from wood fiber mixed with linen or cotton. Countries that portray trees on their bills include India, Suriname, the United Kingdom, Canada, Central African Republic, and Australia. Some countries represent their country on bills with a particular species of tree. For example, Lebanese banknotes—in particular the 5-piastre and 5,000-livre notes—contain as part of their design the cedar tree, which once forested the country but is now restricted to a few protected areas due to overharvest. As recently as 1996, the country of Ghana, at one time the world's largest producer of cacao, featured the harvest of a cacao crop on its 1,000-cedi banknote. Even the United States features trees on its major bills, as frames for buildings of national significance: the Treasury Building ($10), the White House ($20), the Capitol ($50), and Independence Hall ($100).

Stamps provide a similar canvas. Recently, one of my students, Benjamin Bell, visited Olympia's sole stamp-collecting store, seeking information on the portrayal of trees on stamps. The philatelist behind the counter, Christopher Dahle, was an informed and informative member of the American Topical Association, an organization that has served members in ninety countries for over fifty years. They create indices and write articles on the diverse subjects that are displayed, and share that infor-

24. More than eighty countries display trees or tree parts on their stamps.

mation with other members as well as walk-ins such as Benjamin. For three hours, Christopher and Benjamin bent over glassine envelopes of stamps, totting up the countries that portray trees on their postage. Their conclusion? As of 2006, trees and their flowers and fruits appear on the postage stamps of eighty-three countries. The most common tree is the palm, which is featured on the stamps of forty-six countries, from Jamaica to Vatican City. Sixty-two other species of trees—from pines to pomegranates—can be found on stamps from around the world.

A last important category of symbolic expression is that of national flags. Trees or parts of trees appear on at least thirty-six flags. In 1965, Canada adopted a design featuring a stylized red maple leaf on white between two vertical bars of red. The cultural significance of this symbol began with Canada's native peoples, who used maple tree sap as food. Norfolk Island, a self-governing territory of Australia, adopted its flag in 1980; it features a green Norfolk Island pine on a white field between two vertical bars of green, symbolizing the abundant vegetation. The flag of Belize features that tropical country's coat of arms, which includes a mahogany tree; tools of the Belizean timber industry—the squaring axe, the beating axe, and the saw; and the national motto, *Sub Umbra Florero*, Latin for "Under the shade I flourish." In 1943, after gaining independence, Lebanon designed its flag with a green cedar on a white field between two horizontal red bars. Here, as on its currency, the cedar represents immortality, steadiness, and endurance, as referenced in the Bible:

"The righteous flourish like the palm tree, and grow like a cedar in Lebanon" (Psalm 92:12).

Three species are of special significance in all of these identity-affirming representations. The English oak, found across Europe and North Africa, was sacred to the ancient Norse and Celtic peoples. The Celts worshipped their gods in sacred groves of oak trees, erecting temples only much later, when their society had been influenced by the Romans. For them, the oak tree symbolized power and protection. The Greeks also placed high value on oaks, the most revered of which was a great oak tree that grew at Dodona, in northwest Greece. For centuries, pilgrims traveled there to seek guidance. In many cultures, the oak is associated with the gods of thunder and lightning, with fertility, and with such strong and protective figures as the legendary Robin Hood.

The olive tree has long been central to the religions and cultures of the peoples of the Mediterranean. Because its fruits yielded food and oil for light, it was also synonymous with value. In Psalm 128, the olive tree is a symbol of prosperity and divine blessing: "Happy are those who obey the Lord, . . . your sons will be like olive trees around your table." Its branches and leaves were used in coronation ceremonies. And of course, the olive also serves as a symbol of peace and hope, as described in the biblical story of Noah. After the rain finally stopped, Noah sent a dove out as a scout. At first, it returned empty-beaked. "So he waited yet another seven days; and again he sent out the dove from the ark. The dove came to him toward evening, and behold, in her beak was a freshly picked olive leaf. So Noah knew that the water was abated from the earth. Then he waited yet another seven days, and sent out the dove; but she did not return to him again" (Genesis 8:10–12).

A third tree, the palm, is held by many cultures as a central symbol. In Mesopotamia, the world axis passed through a palm tree. The coconut palm produces fruits that were essential to both the nourishment and the religious life of people in India. The coconut was extensively used in Hindu rituals as a replica for the human head. It is also associated with Shiva, because its three "eyes" symbolize the eyes of Shiva.

LIFE TRANSITIONS

What you want for it, you'd want
for a child; that she take hold;
that her roots find home in stony

winter soil; that she take seasons
in stride, seasons that shape and
reshape her; that like a dancer's,

her limbs grow pliant, graceful,
and surprising; that she know
in her branchings, to seek balance;

that she know when to flower, when
to wait for the returns; that she turn
to a giving sun; that she know

fruit as it ripens; that what's lost
to her will be replaced; that early
summer afternoons, a full blossoming

tree, she cast lacy shadows, that change
not frighten her, rather that change
meet her embrace; that remembering

her short history, she find her place
in an orchard; that she be her own
orchard; that she outlast you;

that she prepare for the hungry world
(the fallen world, the loony world)
something shapely, useful, new, delicious.

—Gail Mazur, "Young Apple Tree, December"

As trees grow from seed to sapling to mature tree, they become associated with the most significant transformations in human lives: birth, maturation into adulthood, marriage, parenthood, old age, and death. People of most cultures hold ceremonies to mark these passages, many of which involve trees.

In Sierra Leone, for instance, on the eighth day after birth, a naming ceremony is held. As part of the ritual, an elder woman from the husband's family places a bit of kola nut and alligator pepper in the baby's mouth, to bring courage and long life. In many countries, trees are planted to commemorate the birth of a baby, the life of the child becoming bound with the life of the tree. In Israel, it is customary to plant a tree when a child is born—cedar for a boy, cypress for a girl. In past times, when the child later married, the wood from the birth tree was used to make the *chuppah* (wedding canopy), symbolizing the joining of the families of the bride and groom. Among the M'Bengas in western Africa, when twins are born, the parents plant two trees of the same kind. In New Zealand, the Maori bury the umbilical cord in a sacred place and plant a young sapling over it. As the tree grows, it becomes a *tohu oranga*, or sign of life, for the child. Trees are also traditionally planted at the metaphorical birth of a new historical era. In France after the Revolution (1789–1792), sixty thousand "Trees of Liberty" were placed in the ground, and Belgians planted trees to celebrate their independence in 1830.

For many people, the task of leaving the birth family and moving into independent adulthood is complicated, especially when it involves a transfer across cultures. My own father immigrated as a young man to the United States to attend graduate school, then launched a career in cancer research. He also tended the two-acre lot of garden and trees of our home in Bethesda, Maryland. Although he was a stern parent and ruled our home with iron discipline, I remember the care he showed when we transplanted young saplings from one part of our yard to another, a near-reverent gentleness that I seldom saw in him otherwise. He always made sure that the sphere of soil surrounding a tree's young roots was big enough to absorb the shock of uprooting, and he unfailingly watered it right afterward, to welcome it to its new environment. I liked to pat the dark soil, his big handprint surrounding my little one beside the slender brown trunk that upheld the pliant limbs, so like mine, ready to grow. I have wondered if he saw himself in those transplanted trees, moved abruptly from a small village in India to the frenetic pace of urban America. Just as he had to put down new roots, so did the tree have to connect with its new home.

Marriage is another transition that is often marked with arboreal symbolism. In choosing to marry, each partner brings a unique family history to the other. At my wedding, my mother-in-law presented us with a handmade wedding sampler; the design was in the shape of a tree, with our names and wedding date aligned with the trunk of the tree, and the names of our parents and grandparents embroidered in the intertwining branches. The sampler has hung in our bedroom for twenty-four years and greets me each morning, a reminder of the connections across generations that reinforce our marriage.

Other countries and cultures bring trees into the wedding ceremony and beyond it. For example, in Bermuda, wedding cakes are topped by a tiny cedar tree in a pot, which the couple later plants so that it will grow along with their love. In Czechoslovakia, the friends of the bride plant a tree in her yard and decorate it with ribbons and brightly painted eggs. Among the Ewe of Ghana, when a man desires to wed a girl, his parents visit her parents to propose the marriage. If their proposal is accepted, the suitor sends a present of native cloths and nuts of the kola tree, at which time a day is appointed for the wedding. In Bulgaria, tradition holds that the husband-to-be's best man must make a wedding banner, the pole of which must be from a fruit-bearing tree and chopped from the tree with a single axe stroke. Trees symbolically enter the partnering sphere in longer-term ways; in Czechoslovakia, Russia, and the United States, for example, some newlyweds give away "wedding tree" seedlings to guests as a sign of appreciation and as a hopeful symbol that the marriage will grow steadily in love.

For many of us, trees are potent symbols of longevity and life, as well as of renewal. Evergreen trees, in particular, connote immortality because they always look foliated—though in fact they are constantly shedding and growing new needles. (Douglas-fir needles, for example, live an average of four to seven years before they are dropped.) Deciduous trees, meanwhile, which lose their leaves in fall and sprout new foliage in spring, often are taken to represent rebirth, and trees' seeds and fruits suggest not just fertility, but also immortality. In Norse myth, for example, Idun

("She Who Renews"), the goddess of eternal youth, grows the magic apples of immortality that keep the gods young. In Taoist tradition, the divine peach *shou*, represented by the Chinese ideogram meaning "long life," bestows immortality. In Judeo-Christian mythology, the Tree of Heaven is the source of the primordial rivers that water the earth—analogous to the Tuba tree of the Qur'an, from whose roots spring milk, honey, and wine, all that nourishes and enlivens us.

Trees symbolize the intertwining of life with death. On a family vacation when I was a child, we visited the Wye Oak, which resided on Maryland's Eastern Shore in a state park established expressly for it. During our visit, my father explained that the tree, a white oak that served as the honorary state tree, had already been living there long before the Pilgrims arrived on the continent. It made us feel solemn, as if we should be dressed in better clothes and speaking in hushed voices. Recently a friend sent me a newspaper clipping reporting that after 450 years of life, the famous giant had been toppled by high winds during a thunderstorm. I felt as if an ancient aunt had died, and experienced sadness at losing the grace and wisdom that old age confers, whether of human or tree.

Some trees are used to symbolize death itself. The cypress, for example, symbolizes the finality of death because once cut, it never resprouts from its stump. In Britain, the yew is often found in graveyards, for its roots were believed to soak up the spirits of the dead, then release them from its branches. Indeed, this symbolism is supported by the tree's biology; the whole plant (except for its fleshy fruit) is highly toxic due to the alklaloid taxine. Symptoms of taxine poisoning include vomiting and malfunction of the cardiovascular system, and without treatment death is assured. This plant found its way into the poisonous brew in Shakespeare's *Macbeth:* "Gall of goats and slips of yew, / Sliver'd in the moon's eclipse."

Many cultures incorporate trees into their death rituals. Some of the most common practices include suspending bodies of the deceased from trees, planting a memorial tree, and burying cremation ashes in the hole of a newly planted tree. The custom of "aerial sepulture"—tree and scaffold burial—has been extensively practiced among a wide variety of Native American groups, and documented from the late seventeenth century up to the present time. If trees were present in the area, the dead were placed in the tops, on limbs sufficiently horizontal to support the body. If such limbs—or trees themselves—were absent, scaffolds were used. George P. Belden, the "White Chief"—a frontier trapper and hunter who lived among the Yanktonais, a subtribe of the Dakotas in what is now Minnesota—described this practice from the 1870s:

> These scaffolds are 7 to 8 feet high, 10 feet long, and 4 or 5 wide. Four stout posts, with forked ends, are first set firmly in the ground, and then in the forks are laid cross and side poles, on which is made a flooring of small poles. The body is then carefully wrapped, so as to make it watertight, and laid to rest on the poles. The reason why Indians bury in the open air instead of under the ground is for the purpose of protecting their dead from wild animals. In new countries, where wolves and bears are numerous, a dead body will be dug up and devoured, though it be put many feet under the ground. I noticed many little buckets and baskets hanging on the scaffolds. These had contained food and drink for the dead. The body was left from four to six months to decompose. When this was completed, the bones were given to the family and the remaining flesh was burned.

Planting trees to memorialize the dead has gained popularity in recent times. Many spiritual community leaders purposefully inject life into ceremonies for the dead. For example, Rabbi Arthur Gross-Schaefer of the Community Shul of Montecito and Santa Barbara, California, presides over a ritual that combines the planting of trees with periods of prayer and silent reflection. By nurturing the living tree, family and friends can continue to share memories and stories of the person who has died. Many cemeteries protect old trees, which bring palpable comfort to those suffering loss. When a person in the Yoruba tribe of Nigeria dies, professional mourners are hired to move the real mourners into a condition of frenzied grief. The professional mourner sings in a mournful tone that rises and falls, uttering poetic sentiments: "He is gone, the lion of a man. He was not a sapling, or a bush, to be torn out of the earth, but a tree—

a tree to brave the hurricane; a spreading tree, under which the hearts of his family could rest in peace."

The loss of an innocent victim inspires a special sort of grief. Many institutions now plant trees to commemorate the victims of violent crimes or attacks. For example, the Tarrant County Coalition of Crime Victim Services in Fort Worth, Texas, spearheaded a tree-planting ceremony for National Crime Victims' Rights Week, which occurs each year in April. Other communities around the nation have similar observances, sponsored by agencies such as the Plymouth County District Attorney's Office in Brockton, Massachusetts; the Third Judicial District Attorney's Office in Las Cruces, New Mexico; and the Child Advocacy Center of Central Susquehanna Valley in Northumberland, Pennsylvania. The events of September 11, 2001, led to the dedication of the Victims Tree, a thirty-six-foot-tall evergreen that was decorated with three thousand white lights and three thousand angels, with a victim's name written on each angel. Similar memorial tree groves have been planted in Virginia, New York, and Pennsylvania. In a poem called "Fern-Leafed Beech," Moyra Caldecott expressed the comfort that a human can derive from a tree when death is close by:

> This tree listened
> when my husband died.
>
> I leaned my head
> against its trunk
> and cried.
>
> No words passed,
> but I took its strength
> and knew
> that life at last
> secretly transforms
> until what is seen
> becomes unseen,
> and what has been
> is still to be.

TREES AS METAPHORS OF COMPLEXITY

Trees are one of many mirrors humans have used to better understand themselves. Imagery of trees and forests permeates our poetry, literature, films, and visual art because of their familiarity, their ubiquity, and their ability to represent our needs and psychological complexity. W. H. Hudson's romantic novel *Green Mansions*, which describes the treetop life of the ethereal Rima the Bird Girl, shows us the rainforest as a place of romance and mystery. J. R. R. Tolkien's *Lord of the Rings* trilogy gives us Ents, or "tree shepherds," benevolent treelike beings that join the humans in their fight against evil forces. In the film version, images of the ingenuous humanesque hobbits perched on the shoulders of the gigantic, slow-moving, deep-voiced Ents as they stride to do battle with the evil Saruman and his fearsome Orcs beautifully capture the way people have long viewed trees as protectors.

Fairy tales are often the first stories we hear as children, and some of them are notorious for the frightening way they present the world, especially forests. The Brothers Grimm published collections of folktales that portrayed forests as wild, dark places, full of peril. The American psychologist Bruno Bettelheim interpreted the forest as the "dark, hidden, near-impenetrable worth of our unconscious" and asserted that "being lost in the woods is an ancient symbol for the need to find oneself." He suggested that children be allowed to interpret these tales in their own way, which will enable them to make more personal meaning in their lives. Little Red Riding Hood's walk through the woods reveals in miniature the complexity of real life—not only its dangers, but also its joys (as seen in her anticipation of visiting her grandmother).

As we have seen throughout this book, poetry, too, is rife with arboreal imagery. When, several years ago, I began collecting poems about trees, I sent out an "All Faculty" message requesting input. I received a torrent of responses, not only poems, but also the titles of stories, novels, and films, and even some jokes that have to do with trees. Having collected all of this material, I needed to organize it somehow, and to pull out themes on the relationships between trees and humans. The solution came from Jade Leone Blackwater, the quiet, rather intense student who collected tree logos. She took to heart W. H. Auden's line "Poetry is clear thinking about mixed feelings" and, drawing on the hundreds of poems we collected, identified a number of contrasting themes that can be extended to humans. Three such pairs proved especially prominent: strength and fragility, stasis and dynamism, and threat and protection.

STRENGTH AND FRAGILITY

Trees are strong. They withstand hurricane-force winds. In cities, they silently endure the invective of pollution, soil compaction, and genetic isolation from others of their species. In a poem called "Arbutus, on Galiano," the Canadian poet Pam Galloway finds solace from grief in the lessons of a madrone tree:

> This tree stands
> like a fork of lightning
> grabbed by the earth, its huge vee
> shouting to me
> of all that I could hold, look: the entire sky
> if I would open up my arms, stretch
> if I would let the air smooth my skin,
> let it peel, knowing
> there are stronger layers beneath.

Trees are also fragile. They can succumb to the toxin made by a colony of microbes, the tiny mandibles of bark beetles, or the biting teeth of a chainsaw. A tropical fig species may well go extinct if humans pump enough carbon dioxide into the atmosphere to raise the global temperature a single degree, or if the population of its tiny wasp pollinator is depressed because of the fragmentation of surrounding forests. This same fragility is found in humans: Aristophanes wrote, "Mankind, fleet of life, like tree leaves, weak creatures of clay, unsubstantial as shadows, wingless, ephemeral, wretched, mortal, and dreamlike." His words bring to mind the horrific scene in Francis Ford Coppola's film *Apocalypse Now* in which American planes jettison napalm on a village surrounded by trees. The tall, strong palms instantaneously burst into flames, while the villagers, losing their protective cover, scatter like leaves themselves, powerless to combat this force.

In contrast, trees can show remarkable strength because of their resilience—their ability to persist in the face of intrusion or disturbance by outside forces. In a poem by the Mayan writer Pancho Ernantes, for example, although human actions take their toll on a tree, it is not powerless. Rather, it musters strength to stand and flourish, just as humans can flourish despite being challenged.

I Am a Peach Tree

I am a peach tree.
I used to be little.
The sheep used to eat my leaves.
I had a very hard time growing up
Because the sheep would come so
Often to see me.

Now the boys bother me,
They keep cutting my trunk and my branches.
They shake down my flowers.
When I produce fruits,
They don't wait for them to ripen,
They just eat them green.
That's how they bother me.
Since I grew up alone,
All my companions around me seem different.
When the wind blows hard,
My flowers and little fruits are blown away.
But the rain cares for me, so I will grow.

The poet Sandra Cisneros, in her novel *The House on Mango Street*, describes both the strength and fragility of trees at once in a chapter called "Four Skinny Trees": "Their strength is secret. They send ferocious roots beneath the ground. They grow up and they grow down and grab the earth between their hairy toes and bite the sky with their violent teeth and never quit their anger. This is how they keep." We humans exemplify a similar strength and fragility. Our bodies allow us to perform amazing feats—ascending into the canopies of trees, climbing mountains, running marathons. But we are fragile, too. Too often, I rely on the opinion of others to define my sense of self. And although I have given birth to two children, the knowledge that I can protect them from almost nothing makes me vulnerable to every danger, real or imagined, in the world. Whenever I climb a strong-limbed, thick-barked tree that exposes tender new foliage, I see myself in each contradictory limb.

STASIS AND DYNAMISM

The word *tree* is derived from the word *dūrus*, the original Proto-Indo-European term meaning "solid, tree." It later emerged as the Old Norse *tre*, Persian *dar*, Hittite *taru*, Serbian *drevo*, Greek *drys*, and Sanskrit *dāru*, all meaning "wood." In English, the term evolved into both *tree* and *true*, sharing roots with the words *endure* and *durable* and meaning with the notions of hardness and continuance. And indeed, trees exhibit characteristics of stability, constancy, and permanence, being rooted in place, having extremely long lives relative to humans, and reacting slowly to most environmental stimuli. Yet trees are also highly mobile, moving about in space. As John Muir observed, "I never saw a discontented tree. They grip the ground as though they liked it, and though fast rooted they travel about as far as we do. They go wandering forth in all directions with every wind, going and coming like ourselves, traveling with us around the sun two million miles a day, and through space heaven knows how fast and far!"

Trees can be wildly dynamic, not only as their branches are pushed around by the wind, but also in the way their leaves are shed and rebirthed each year, their limbs drop and trunks topple, logs ride down rivers, and their nutrients enter circulatory cycles of the ecosystem, translating them into more mobile forest inhabitants. In her poem "I Was Sleeping Where the Black Oaks Move," Louise Erdrich describes a dramatic flood event that, when the waters subside, brings about continuity in transformation.

> We watched from the house
> as the river grew, helpless
> and terrible in its unfamiliar body.
> Wrestling everything into it,
> the water wrapped around trees
> until their life-hold was broken.
> they went down, one by one,
> and the river dragged off their covering.
>
> Nests of the herons, roots washed to bones,
> snags of soaked bark on the shoreline:
> a whole forest pulled through the teeth
> of the spillway. Trees surfacing
> singly, where the river poured off
> into arteries for fields below the reservation.
>
> When at last it was over, the long removal,
> they had all become the same dry wood.
> We walked among them, the branches
> whitening in the raw sun.
> Above us drifted herons,
> alone, hoarse-voiced, broken,
> settling their beaks among the hollows.
>
> Grandpa said, *These are the ghosts of the tree people,*
> *moving above us, unable to take their rest.*
>
> Sometimes now, we dream our way back to the heron dance.
> Their long wings are bending the air
> into circles through which they fall.
> They rise again in shifting wheels.
> How long must we live in the broken figures
> their necks make, narrowing the sky.

THREAT AND PROTECTION

Being at once threatening and protective seems the ultimate contradiction, but trees symbolize both in human expression. As children, my siblings and I watched *The Wizard of Oz* on our black-and-white television each year. My brothers and sisters screamed at the giant talking head of the automated Wizard, flames roiling and tympani crashing. I, however, was more profoundly shaken by the humanized apple trees with their sullen, bark-embedded faces and gnarly arms that threw apples—hard!—at the innocent travelers. The idea of trees as spiteful and angry and out to hurt people was completely counter to my own view. Then there are films like *Aguirre, Wrath of God* and *Apocalypse Now,* based on Joseph Conrad's novel *Heart of Darkness,* which link the journey from city to jungle with the ethical journey from morality to evil. In both, the rainforest becomes a character of its own, whose dark mysteries sunder the protagonists' link to sanity.

Yet trees and forests also provide protective, nurturing symbolism for humans. Starting with the early Tarzan movies, and progressing to more recent films for young people such as *The Blue Lagoon* and *Fern Gully,* trees and forests symbolize Edenic places where innocence and goodness triumph over the evils of civilization. In Harper Lee's *To Kill a Mockingbird,* a neighborhood tree serves as a spot where Scout and her friends leave special tokens for each other, becoming a childhood refuge in the face of the confusing, risky world of grown-ups.

As providers of the basics of shelter and food, trees are also symbolic protectors in more mundane ways as well. When I read Johan Wyss's *Swiss Family Robinson,* about a family cast upon a tropical desert island who survived nicely on what nature provided, I dearly wished to get shipwrecked on the same island so I could move into Falconhurst, the multistory tree house that they constructed in the spreading crown of a fig tree. Jean George's *My Side of the Mountain,* the story of young Sam Gribley, who leaves the city to live off the land in the forests of New York State, evoked fantasies of striking out for the Catskills with nothing but a flint and steel in my pocket. In *Little House in the Big Woods,* Laura Ingalls Wilder depicts the hard but rewarding life that a pioneer family ekes from the deep forest around their small clearing. Films, too, explore how trees give the hope of health to humans. For example, in *Medicine Man,* the protagonist climbs into the forest canopy to secure plants that will produce an anticancer compound. Tragically, his discovery comes just as timber cutters enter the scene, making humans—all too painfully—into agents who can negate the protective capacity of trees.

All of these human expressions contribute to a collective consciousness of the sometimes contradictory ways in which trees and humans interact. Wherever I look, whether in movie lines or memorial services, on our flags and logos and currency, in our poems and films, I see that trees enrich and instruct our lives in ways at once simple and complex.

Chapter Eight

SPIRITUALITY AND RELIGION

God is the experience of looking at a tree and saying, "Ah!"

—*Joseph Campbell and Bill Moyers,* The Power of Myth

A giant *Ceiba pentandra* tree marked the spiritual beginnings of my marriage to Jack Longino, now my husband of twenty-four years. As graduate students, we had met in the rainforest of Costa Rica and fallen in love. We looked forward to exchanging our vows in the hybrid Hindu/Jewish/Presbyterian ceremony we were planning for our family and friends. But it seemed appropriate to seal our commitment in the crown of a rainforest tree, a hundred feet off the ground. After pulling ourselves aloft with ropes and climbing gear, Jack and I settled into the broad branches of the ceiba for the afternoon. As we toasted our decision to spend the rest of our lives together, we both observed that two smaller, separate branches farther out the limb had connected to the single larger, stronger branch on which we were sitting. There, held in the strong limbs of the tree with the man I loved, looking over the rainforest roof, I felt connected to something beyond myself, beyond the two of us—to something spiritual.

Trees satisfy the penultimate step of the modified Maslow pyramid, the need for spirituality. Because of their iconic form, their persistence through time, their rootedness in the soil, their skyward-extending branches, trees constitute for us a connection between Earth and the heav-

25. *A giant silk-cotton, or ceiba, tree is a common sight on the tropical skyline. These trees grow remarkably quickly, creating broad, sweeping crowns. Photo by Greg and Mary Beth Dimijian.*

ens that is both physical and spiritual. In important ways, they allow us to expand our sensibility beyond the mundane.

WHAT IS SPIRITUALITY?

Although the terms *spirituality* and *religion* are often used interchangeably, they carry different meanings. Religion describes a set of beliefs about the supernatural or divine that answers questions of origin, purpose, morality, and mortality for its followers. In contrast, spirituality is defined in *Webster's Third* as the animating force traditionally believed to be within living beings; the part of a human associated with the mind and feelings as distinguished from the physical body. The Dalai Lama puts it like this: "Spirituality guides people about contentment, timelessness, right and wrong, self-discipline, change, peace, sharing, forgiveness, and tolerance."

Both religion and spirituality help people answer the fundamental questions of life: Who am I? Where did I come from? To what forces am I connected? Will it matter that I lived? Some people respond to these questions by joining a particular religion and aligning themselves with its creeds. Others find the answers in the patterns and activities of the world around them. Many have looked to nature—and specifically to trees—for guidance in living a more purposeful and meaningful life. Sogyal Rinpoche, the Buddhist author of *The Tibetan Book of Living and Dying*, wrote the following:

> Think of a tree. When you think of a tree, you tend to think of a distinctly defined object. But when you look at it more closely, you will see that it has no independent existence. When you contemplate it, you will find that it dissolves into an extremely subtle net of relationships that stretch across the universe. The rain that falls on its leaves, the wind that sways it, the soil that nourishes and sustains it all the seasons all form part of the tree. As you think about the tree more and more you will discover that everything in the universe helps make the tree what it is; that it cannot be isolated from anything else and at every moment its nature is subtly changing.

The word *human* could easily replace the word *tree* in this piece, and that can start us on our way to thinking about trees, spirituality, and humans.

In my life, spirituality arose from a range of traditions. My father—who died in 1999—was a Hindu; my mother is Jewish; and for several years, my siblings and I attended a Unitarian Sunday school. I conduct fieldwork in the Costa Rican village of Monteverde, founded by North American Quakers in 1950, and I attend the Friends Meeting whenever I am there. From this motley set of experiences, I have come to understand that beliefs are held in common across many faiths, and a number of basic tenets are linked to nature or understood through metaphors of nature. In recent years, I have abstracted these personal experiences to explore how people of different faiths understand trees and forests—as expressed in their holy texts, their religious practices, and the architecture of their places of worship.

In 2001, with the support of a Guggenheim Fellowship, I investigated how scientists might better communicate science and conservation to non-scientific audiences, especially those whose bent is not to visit a science museum or read a natural history magazine. One aspect of this work was to present my thoughts on the connections between trees and spirituality in places of worship.

Before taking the pulpit, I attended services, listening to the sermon, singing the prescribed hymns, and contributing to the collection plate. After several months, I approached individual clergy, rabbis, and priests to offer to give a sermon or lead a discussion group on trees and spirituality. The congregations I addressed ranged from fundamentalist to progressive and included Episcopalians, Baptists, Unitarian Universalists, Zen Buddhists, Reform Jews, Conservative Jews, Roman Catholics, Methodists, and secular interfaith organizations. I presented my talks not as a religious person, but as an ecologist interested in understanding trees with my intellect and as a human being who cares deeply about trees with my heart. My audiences were eager to share their own experiences. After the sermons, members of the congregation would scribble quotations relating to forests from their holy scriptures in my notebook, sing a hymn that mentions trees, or offer an introduction to other congregations.

On one occasion, I spoke from the *bima* (meaning "high place," the raised platform from which the Torah is read) of the Jewish synagogue in Olympia, Washington. About fifty congregants had come on a cold, wet January night—a good night to stay home—to hear about trees, spirituality, and Judaism. One man sat in the very back row. He was elderly and blind, and everything he owned appeared to be resting damply in a shopping bag beside him. After the discussion, he stood up and directed his unseeing eyes upward. "When it is cold and raining, like tonight," he said, "and I stand under a tree, I stay dryer and warmer than when I am out in the open. Trees protect me." He paused. "Sort of like God." No conservationist could have put it better. Trees hold a central place for humans in the realm of the spiritual and religious.

TREES AS CONNECTIONS TO DIVINITY

COSMOLOGY AND THE WORLD TREE

Trees have long provided a useful metaphor for explaining our origins. A "world tree" is central to many creation myths and symbolizes the ultimate reservoir for the forces of life that continually regenerate existence. It frequently represents the axis of the universe—the *axis mundi*—that connects different realms of the cosmos. Its branches hold up the heavens, its trunk stands in the earthly realm, and its roots descend into the underworld. In Norse mythology, the ash tree Yggdrasil—resilient, supple, and strong—connected the three major realms of all beings: the upper realm, occupied by gods; the middle realm, occupied by humanity; and the lower realm of the dead.

Other cultures employ similar imagery. The world tree of many of the cultures of pre-Columbian Mesoamerica—in particular the rich Mayan civilization—represented the concept that the central core of the world is a tree. In the Yucatán Book of Chilam Balam of Chumayel, which contains much of what the Indians remembered of their old culture after the Spanish Conquest, this world tree is a pervasive symbol of the creation and ordering of the world. It is the axis of Earth-Sky, with its roots lying in Xibalba, the Underworld, its top reaching into the heavens. It is most often depicted as the silk-cotton tree (the tree in which my husband and I married ourselves), striking in both stature and structure. Some species can grow to 220 feet or more in height, with a straight, branchless trunk that culminates in a huge, spreading canopy and "buttress" roots that can be taller than a man—an appearance that reinforces its central presence in the cosmology of this civilization.

Trees occupy this same keystone realm in other cultures. The ancient Egyptians believed that a holy sycamore grew at the threshold between life and death. According to the Book of the Dead, twin sycamores stood at the eastern gate of heaven, from which the sun god Ra emerged each morning. The sycamore was also regarded as a manifestation of the goddesses Nut, Isis, and especially Hathor, who was called Lady of the Sycamore. The tree was often planted near tombs, and burial in coffins made of sycamore wood returned the dead person to the womb of the mother tree goddess. In the Shinto religion, sakaki, an evergreen tree in the camellia or tea family, is sacred. The *Kojiki* (Record of Ancient Matters), a text dating to the eighth century, describes how this tree brought light and life to a darkened world. In the Grand Shrine of Ise, the foremost Shinto shrine in Japan, the sakaki tree is represented by a sacred central post around which the wooden shrine is built. Branches of sakaki are used in rituals, often with paper streamers or mirrors attached, as offerings.

In the Hindu tradition, too, the world tree—a banyan—is a central image, but it is turned upside down, being rooted in the heavens and bearing its fruit on earth. All the gods and goddesses, all the elements and cosmic principles, are its branches, but each one is rooted in Brahman, who is identified with the stem of the sacred tree itself. As with the ceiba tree of the Mayans, the structure of the banyan tree reflects the spiritual concept of uniting heaven and earth. The banyan spreads over huge areas by sending aerial roots down from its branches. When the aerial roots touch the ground, they themselves take root and develop into stems, so that a single tree can constitute a whole forest—symbolizing the idea of multiple gods and spirits in all their localized aspects being part of the one ultimate source. Like Hindu cosmology, the medieval doctrine of Jewish mysticism known as Kabbalah envisioned a world tree with roots in the world of the spirit (the unseen) and branches upon the earth (the seen). Kabbalists used this tree (*Etz ha-Chayim* in Hebrew) to understand the nature of God and the manner in which the world was created out of nothing. The branched candlestick known as the menorah, one of the most ancient symbols of Judaism and used by Jews today to celebrate Chanukah, the Festival of Light, has links to this Hebrew Tree of Life. In imagery, this tree offers a "map" of Creation.

The notion of a tree as the *axis mundi* is not restricted to heavily forested lands. The Sami people (also known as Saami, Lapps, or Laplanders) traditionally conceived of the world as a cosmic tent, with different levels of reality that corresponded closely to the different worlds of Norse mythology. In some Siberian tribes, such as the Yakut of southeastern Siberia, the connection between worlds is the cosmic tent-pole; the cosmic smoke-hole at the top is the North Star, and humans may gain access to other worlds by climbing the world tree.

Many of these same cultures connected the world tree—which linked the worlds of gods and humans—with the provider of things that sustain life, and thence to the symbolic Tree of Life. In Babylonian mythology, the Tree of Life was a magical tree that grew in the center of Paradise, with primordial waters flowing from its roots. In Egyptian mythology, the sycamore tree on which the gods sat bore nourishing fruits that fed the blessed. Many species of ash—the species of tree that was Yggdrasil—exude a sugary substance that the Greeks called *me'li*, or honey. This may explain how Yggdrasil could rain honey on the world, while mead—wine made with honey—flowed from its branches.

The materials provided by these sacred trees could be symbolic as well as physical. In ancient India, the leaves of the world tree were called Veda (knowledge), since wisdom and understanding were considered to be the essential fuel for spiritual growth. The image of trees bearing leaves that become accessible to those associated with this symbolic tree permeates many Eastern cultures. Zoroaster said, "To the soul, it [the Tree] is the way to heaven." Proverbs 3:18 says, "It is a tree of life to those who lay hold of it; those who hold it fast are called blessed"—"it" being wisdom, which can be equated with God's revelation of himself to believers.

Of course, we all know the story of the Garden of Eden: "The Lord God made to grow every tree that is pleasant to the sight and good for food, the tree of life also in the midst of the garden, and the tree of the knowledge of good and evil." When he created Adam, he put him in the garden "to till it and keep it. And the Lord God commanded the man, saying, 'You may freely eat of every tree of the garden; but of the tree of the knowledge of good and evil you shall not eat, for in the day that you eat of it you shall die'" (Genesis 2:9, 16–18). John McLain, a divinity school graduate turned academic grant-writer and a friend of mine, calls the Tree of Life "a symbol of the goodness of creation: its sustenance, shelter, and beauty." Of the Tree of the Knowledge of Good and Evil, however, he asks:

> Why were Adam and Eve allowed to eat from the other trees in the garden but not this one? Perhaps knowledge of good and evil makes us like God, but in all the wrong ways. Like trees, we ought to simply grow where we take root and consume only that which is given to us. The forbidden fruit opens a Pandora's Box of possibilities for humans who cannot yet wield its choices responsibly. The story, however, holds its own seeds of hope. Adam and Eve, though expelled forever from the Garden, also set humanity on a path to restore its relationship with God and regain its place in creation.

What about the spiritual roots of cultures that evolved in the landscapes of my own homeland, North America? Many of the ways that the symbolism and structure of trees have infused the cosmology of the peoples I described above are also evident in the rich cultures of Native Americans. It is important to recognize that the North American continent is home to over five hundred indigenous tribal peoples, each with their own spiritual beliefs and rituals, not to mention natural environments, so to speak of a single, monolithic "Native American spirituality" is incorrect. However, it is true that for many indigenous peoples, spirituality that is directly linked to nature played and continues to play a central role in their lives. Chief Seattle, a nineteenth-century leader of the Salishan people of what is now Washington State, said that "all things share the same breath—the beast, the tree, the man, the air shares its spirit with all the life it supports." The Native American historian Angie Debo, in generalizing about Native American attitudes, described the archetypal Indian as "deeply religious. The familiar shapes of earth, the changing sky, the trees and other plants, the wild animals he knew, were joined with his own spirit in mystical communion. The powers of nature, the per-

26. The biblical Tree of Life is a central figure in religious and cultural imagery worldwide. Erastus Salisbury Field, The Garden of Eden, *photograph © 2008 Museum of Fine Arts, Boston.*

sonal quest of the soul, the acts of daily life, the solidarity of the tribe—all were religious, and were sustained by dance and ritual."

The relationship between the people and the land was one of interdependence, and the form of worship varied with the environment. To those who relied on buffalo for food, clothing, shelter, and implements, for example, the buffalo played a central role in their cosmology. Those who dwelt in the Pacific Northwest considered the forest and certain trees, such as western red cedar, fundamental to their spiritual lives. The practical uses intertwined with the ritualistic and spiritual applications. In such a world, everything is imbued with spirit, and there is a constant dialogue among all the manifestations of creation. One theme found in some tribes explains the universe as being composed of multiple layers that are linked by the symbolic world tree, which has its roots in the underworld, stretches through the world of humans and animals, and has its crown in the sky world. All these worlds function together in a cosmic whole. World order is founded on a balance of interrelationships between humankind, the universe, and the supernatural powers. Native Americans see it as their task to live in harmony with this universe. Lorain Fox Davis of the Cree-Blackfeet tribes, director of the Rediscovery Four Corners, a nonprofit organization that serves Native American youth and elders, described the Sun Dance of the Plains Indians as one example of this focused spiritual expression.

> It is a ceremony of sacrifice and thanksgiving honoring the sacredness of the circle of life. From sunup to sundown, each day for four days they dance and fast—without food, or water. Each day the four major races of people are prayed for; children, adolescents, adults and elders are prayed for; those who swim, fly, crawl, the green and growing things, and the stone people are prayed for; each of the four sacred directions, the powers of those directions, and the elements are prayed for. Everything is brought together in the circle, all living things are danced and sung for. In the center of the circle is the Tree of Life, and the people dance around her. They dance and sing and focus on "all our relations, and our humble place in the circle of life." For four days the dancers pray for all of creation first, before they include themselves. They end all prayers with "O MITAKUYE OYASIN," meaning "I do this for all my relations (or all sentient beings)."

TREES AS GUIDEPOSTS TO ENLIGHTENMENT

Trees often serve as guideposts in the human search for meaning and direction. In the introduction to this book, I described the experience of bringing people who had never seen a forest into the forest canopy. One of these was Brian Arululak, an Inuit artist from the tundra of Nunavut, Canada. A man of quiet temperament, he was skilled at creating small paintings. For the first five days of our camping retreat our tents nestled at the base of a tall hemlock tree, which he had learned to identify by species name, *Tsuga heterophylla*. Brian would paint detailed scenes of his native tundra: ice fishing, polar bears, and the flat, treeless landscapes of the far north that he calls home. On the last day, he gave me a painting that contained several tall, skinny trees that looked much like long garden stakes—reminding me that *nabaaqtut*, the word for "tree" in his native

Inuktitut language, actually means "pole." He had also painted a stone cairn in the shape of a T directly in front of the trees he had drawn. When I asked him what the cairn was for, he said, "Where I live, there are no trees for wood fences. We use piles of stones to mark the trails in the tundra. These cairns are our guideposts." He pointed to the tree and continued, "This week, sleeping under our *Tsuga heterophylla*, I have seen that trees are your guideposts."

Brian's insight was correct. Towering over the human form, with lifespans that transcend human generations, trees are often revered as links between heaven and earth, between the spiritual and the mundane. Hermann Hesse wrote: "In a tree's highest boughs, the world rustles, its roots rest in infinity." The same sense is found in the words of Carl Jung, the founder of analytical psychology, who wrote: "Trees in particular were mysterious, and seemed to me direct embodiments of the incomprehensible meaning of life. For that reason, the woods were the place that I felt closest to its deepest meaning and to its awe-inspiring workings."

Many religious stories about spiritual awakening include trees as central figures. One seeker of enlightenment, the Buddha, was born over 2,600 years ago in a grove in Lumbini, in what is now Nepal. The Buddha's mother, Queen Maya, was traveling to her parental home when she gave birth. Standing under a sal tree, she grasped a low branch to support herself, and her infant, Siddhartha, emerged. As a youth, Siddhartha renounced his kingdom to avoid the unending cycles of birth, aging, and death. For many years, he wandered in the wilderness as an ascetic. After years of study and meditation, Siddhartha sat under a large Bodhi tree, or pipal, in the Mahabodhi temple grounds in Bodhgaya (now eastern India), resolving not to rise from the spot until he attained enlightenment. After many days, Siddhartha arose as the Buddha, the Awakened or Enlightened One. He was thirty-five years old. He spent the next forty-five years teaching others the path to enlightenment, mostly in the shade of the great banyan trees in village squares. This poem by Kenneth Rexroth, in which the Buddha addresses one of his principal disciples, Ananda, describes how the Buddha used trees metaphorically in his teachings:

> *The City of the Moon*
>
> Buddha took some Autumn leaves
> In his hand and asked
> Ananda if these were all
> The red leaves there were.
> Ananda answered that it
> Was Autumn and leaves
> Were falling all about them,
> More than could ever
> Be numbered. So Buddha said,
> "I have given you
> a handful of truths. Besides
> these there are many
> thousands of other truths, more
> than can ever be numbered."

At the age of eighty, Buddha knew he was close to death, and so he lay down in the shade of two sal trees, which were in unseasonably full bloom. For hundreds of years after his death, the Bodhi tree represented Buddha's image in place of a human likeness. The tree that stands in the Mahabodhi temple today is a descendant of the tree that was growing in Buddha's time. A cutting of that tree was taken to Sri Lanka in the third century B.C., and a sapling from it was later brought back to Bodhgaya. Since then, the Bodhi tree has been burned and cut down various times, but as legend has it, it has miraculously grown back each time.

I enjoy finding connections in the biology of sacred trees that might explain the spirituality that is attributed to them. Consider, for example, the sal tree, which marked both the birth and death of the Buddha. While continuing to satisfy spiritual needs through its association with Buddhism, it meets a variety of more down-to-earth human needs as well. It is in the family Dipterocarpaceae, native to Southeast Asia, whose members have a tall, strong stature, with erect trunks and excellent wood. Lumber from the sal tree is used for building houses—the pores in its wood contain a resin that makes the timber very durable—and people use the leaves as dinner plates, and as cattle fodder and fertilizer.

The banyan tree, too, mirrors its sacred status in its biology. Since a seedling does not require soil to become established, it can grow perched on another, larger plant or a rock wall. This aboveground position gives the young tree more access to light and less competition from ground-dwelling plants. Water comes only from rain and the humidity of the air, while nutrients such as nitrogen are absorbed directly from rainfall, decomposing leaves, and dead insects. From its elevated nursery, the young banyan sends its roots toward the ground; once the roots reach soil, the plant switches over to growing as a normal tree. The roots of an adult banyan follow cracks and crevices where soil and water accumulate. Not only do these distinct forms of the banyan reflect human physical and spiritual development, but they can also be related to the multiple ways that religion enters the life of a follower.

Other attributes of the banyan mirror the spiritual characteristics of protection and strength, attributes that are taught by leaders of all of the world's religions. It supports a thick overhanging canopy of deep green leaves, which shades the understory plants from intense sunlight. Its leaves are heart shaped, with an elegant tail-like "drip-tip" that guides rainfall down to the soil at a gentle pace, protecting the soil from the pounding monsoonal rain.

SACRED GROVES

Traditions hold that certain forest groves are dwelling places for spirits and places where humans and deities can walk on common ground. In the first century B.C., Seneca, the Roman senator and orator, wrote: "When you enter a grove peopled with ancient trees, higher than the ordinary, and shutting out the sky with their thickly intertwined branches, do not the stately shadows of the wood, the stillness of the place, and the awful gloom of this doomed cavern then strike you with the presence of a deity?"

My father was respectful of trees. When I was a child, I would watch him when he had to periodically trim back the branches of the maple trees that lined our driveway to maintain a clear view of the road outside our house. He would pause and deliberate before each cut he made with the lopper, as he called the plant clippers, trying to minimize the number of limbs he lopped and the length of each shorn segment. When he was finished with his cuts, he would have me fetch him the small can of tree paint from a special shelf in the garage. With great care, he would coat each exposed surface to help heal the wounds he had made.

Perhaps these sensibilities came from his having been born and raised in India, where early inhabitants perceived a godly element at work in places of natural beauty, especially in trees. Centuries ago, many villages set apart sacred land for the "tree spirits," or *vanadevatas.* According to the Hindu Deva Shastra, "trees serve as homes for visiting devas [spirits] who do not manifest in earthly bodies, but live in the fibers of the trunks of the trees, feed from the leaves, and communicate through the tree itself" (verse 117). Would-be parents propitiated the spirits by tying toy cradles to the branches of trees in sacred groves. The groves were sites for celebrations. Although it was permissible to collect deadwood as fuel for cooking fires, doing damage to the sacred grove, especially by felling a tree, could easily invite the wrath of the local deity, causing disease, natural disaster, or the failure of crops.

Despite the ebbs and flows of many political and religious systems, spiritual beliefs have been the prime force preserving these groves in this country of one billion people. In 1995, 13,270 sacred groves existed in India, ranging in size from a few trees to hundreds of acres, often associated with water sources. All of them have a resident deity; often it is Shiva or Vishnu, two of the major Hindu incarnations of God.

The basic elements of nature, in the form of Prithvi (Earth), Agni (Fire), Jal (Water), Vayu (Air), and Akash (Space), have been revered from ancient times as the abodes of God, manifested in the land and water, flora and fauna. As a result, the groves got protection for spiritual as well as cultural, social, and ecological reasons. One especially revered tree species is oak, important for fodder and fuel. It is viewed as improving soil fertility through efficient nutrient cycling and conserving soil moisture through humus buildup and via its deep root system, with root biomass uniformly distributed throughout the soil profile.

In some groves, nothing may be removed, while in other groves, people are allowed to gather materials such as fallen branches and leaves from the forest floor or fruit from the trees. The Western Ghats—a mountain range that runs along India's southwest coast—are one of the world's biodiversity "hotspots," as proclaimed by Conservation International in 1999. Biologists estimate that two thousand plant species and three hundred vertebrate animal species are endemic there—meaning they live in this region and nowhere else. In Kerala, one part of the Western Ghats, some sacred groves are dedicated to the conservation of snakes, which protect agricultural crops by controlling insect and rodent populations. In the arid region of Rajasthan, in northwest India, the Bishnoi tribes manage sacred groves called orans. These provide a protective habitat for the blackbuck, a striking ungulate that lives in open plains and one of the fastest animals on Earth. Adult males have spiraling ringed horns that can reach nearly two feet in length. The Indian gazelle is another species that survives in large part due to protection of the Bishnoi sacred groves. This lovely ungulate has a smooth, glossy summer coat of a warm biscuit color, with dark chestnut stripes on the side of its face. It is a shy animal and avoids humans. The gazelle is of interest because of its ability to survive severe drought, being able to procure sufficient water from eating desert plants and licking dew from foliage. Although its habitat is protected in the groves, the gazelle is not; as a result, it is classified as extremely vulnerable and is included on the highly endangered species list.

Sacred groves are not restricted to India, of course. In Europe, people have sought out the spiritual protection and mystical energy of forests since prehistoric times. Trees and spirituality are intimately linked in druidism, a pre-Christian religion whose adherents worshiped the spirits of trees. The word *druid* (first written down in Greek) is traced to the Indo-European roots *deru-*, meaning "firm, solid, steadfast," and hence

27. Druid religion and culture predate Christianity, and many of their ideas and traditions were adopted by the Christian church. Charles Knight, "Archdruid in His Full Judicial Costume," from Old England: A Pictorial Museum, *1845.*

"tree"; and *-weid,* meaning "to see," and by extension "wisdom" or "knowledge." In pre-Christian Celtic society, druids formed an intellectual class of philosophers, judges, educators, doctors, astronomers, astrologers, and advisers to kings. The oak was their sacred tree, and they venerated mistletoe, which grew in its branches. Nearly every tribe in ancient Gaul possessed a sacred meeting place surrounded and protected by trees in which a local deity was believed to reside. These were centers of religious ritual, and the trees' destruction was viewed with the same horror that would attend the burning of a church or bombing of a mosque today. Cutting down a tree in a sacred grove could mean death for the offender.

Our knowledge today of druidic prayers is limited. Theirs was an oral culture, and their sacred songs, prayers, and rules of divination and magic were learned by heart. The druids were suppressed in Gaul and Britain after the Roman conquests, and a prescript of Augustus forbade Roman citizens from practicing druidic rites. Instruction then became secret, carried on in caves and forests. Gradually, the Christian church absorbed many Celtic practices: pagan gods and goddesses were transformed into

Christian saints, and pagan temple sites became cathedral grounds. By the seventh century, druidism had been driven deep underground throughout most of the formerly Celtic lands, and because its traditions were mainly oral, it nearly died out completely, with little that could be deemed "authentic" remaining. Eleven centuries later, however, England experienced a revival of interest in the druids, and this interest has continued into the present day. Several notable Britons were initiated into druidic orders, including William Blake and Winston Churchill. Modern druids are especially inspired by a concern for the environment; they protect groves that support mistletoe, fight the culling of badgers in the countryside, and disseminate information on global climate change.

Sacred groves exist in many other areas of the world. In Africa, for example, areas of what are now Nigeria and Ghana continue to retain forests because their spiritual properties were considered valuable by those who lived in and around them. A tree or woodland may have become sacred for various reasons: to mark the spot where a founding ancestor stopped, to provide habitat for totemic animals, or because it is linked to a desirable human characteristic. For example, the Osun-Osogbo Sacred Grove of Nigeria, which contains dense forests, is dedicated to the fertility god in Yoruba mythology; dotted with shrines and sculptures, it was designated a UNESCO World Heritage Site in 2005. In the arid lands of West Africa the sausage tree produces large woody fruits that look like enormous elongated hanging bags. Because these represent powerful fertility images, women who are nursing children tie strips of fabric on the tree to ask for numerous offspring. Then there is the tamarind, which is always green and has hard and durable wood. Its severe and imposing appearance leads to its association with the presence of spirits, and people endow it with values related to tenacity under duress.

TREES IN SACRED ARCHITECTURE AND THE SACRED ARCHITECTURE OF TREES

I have always been struck by how, so often, our houses of worship reflect the physical structure of trees in their architecture. People have speculated that Greek temples, with their fluted stone columns, were architectural renditions of sacred groves. The Gothic cathedral reveals the linkage with trees especially strongly, with its ribbed vaults, flying buttresses, columns, and spires. When I enter the nave of a cathedral—whether it is the massive Notre Dame in Paris or a tiny church in a Costa Rican village—my eyes and spirit move upward, just as they do when I walk in tall forests. John Muir may have been feeling this same sort of reverence when he named the Cathedral Grove of coast redwoods, in what is now John Muir Woods National Monument, north of San Francisco. This name is appropriate, as Muir equated his perception of wild trees with a sense of the divine, as expressed in this passage from *My First Summer in the Sierra* (1911): "A few minutes ago every tree was excited, bowing to the roaring storm, waving, swirling, tossing their branches in glorious enthusiasm like worship. But though to the outer ear these trees are now silent, their songs never cease. Every hidden cell is throbbing with music and life, every fiber thrilling like harp strings, while incense is ever flowing from the balsam bells and leaves. No wonder the hills and groves were God's first temples."

Trees commingle with religion and religious architecture not just suggestively, but physically as well. In India, Hindu temples are nearly always shaded by a banyan tree, which is mirrored in an adjacent pool of clear water. Some of the most powerful examples of the intertwining of nature with religion are found at Buddhist temples in Thailand, where the roots and trunks of banyans grow so intimately with the stone walls that the two become indistinguishable. Perhaps the largest religious monument ever built is the 500-acre Angkor Wat in Cambodia, from the early twelfth century, a classic exemplar of Khmer architecture and art. It is dedicated to the Hindu god Vishnu the Preserver. Throughout the temple complex, carvings and sculptures depict gods, battle scenes, dancers, and events in Hindu mythology. Working in sandstone, a fairly soft material, made the construction project easier for the five thousand artisans and fifty thousand laborers who built the temple over the course of thirty years. One of the most striking characteristics is the strangler figs that grow on and between the sandstone blocks, in some cases holding them together like a mesh bag holds mangoes from the market.

Individual trees can serve as a spiritual refuge as well. Banyan trees are still found in almost every village in India, and small shrines are traditionally built under them. They are the heart of the village's sacred life because they encompass the spiritual qualities that humans cherish: longevity, serenity, and resilience. My father often spoke of his youth in the small village of Thane (pronounced TAH-nah), where he and his father and grandfather would sit in the central square in the shade of such a tree. Today, that little village is a bustling suburb of Mumbai, but the square and its banyan tree survive.

Children especially love banyan trees, which have a wonderful form for climbing, with broad, sturdy branches and numerous roots hanging down like thick vines, ideal for swinging. When I was seven years old, my family visited India to see our many relatives. Some of my dearest memories of that holiday include swinging on the roots of the banyan trees in each village we stopped at. Even at that age, I perceived that the spiritual and social lives of the village were entwined. Adjacent to the bough-protected altars to Shiva, Ganesha, and other Hindu deities, the *panchayat*, or village council, would meet. One of the oldest democratic systems in use today, this council makes decisions on key social and economic issues, acting as a conduit between the villagers and their regional government. Traditionally, too, merchants, traders, and the village barber and fortune teller would set up shop under the banyan's spreading branches. In fact, the tree gets its name from the Gujarati word *banian*, which means "trader."

Many other cultures hold the banyan as both a spiritual and social anchor. In Bali, banyan trees are considered "elders" of the tree kingdom and accorded special respect. Motorists honk when they pass a banyan tree on the road—not the usual beep of impatience, but a polite greeting to the tree. In Cambodia, people believe that their history is like that of the banyan tree, composed of a thousand branches intertwined, merging the past and present, forever changing and growing.

28. The banyan tree is a central participant in the social, cultural, religious, and economic life of many villages in India and other Asian countries.

TREES IN SPIRITUAL WRITINGS

The birthplace of Christianity, Islam, and Judaism is the arid desert of the Middle East. In the holy writings and scholarly commentaries of these religions—the Old and New Testaments, the Qur'an, the Talmud—over twenty species of trees, or products of trees, are mentioned: acacia, almond, apple, carob, cedar, citron, cypress, date palm, ebony, fig, frankincense, oak, olive, pine, pistachio, plane, pomegranate, poplar, sycamore, tamarisk, terebinth, walnut, and willow. I wanted to find out how these trees were used by the followers of these faiths, so I downloaded the Old Testament and the Qur'an and did a computer search for the words *tree* and *forest.* I got 328 hits. I then categorized each verse into the different uses to which the trees were put. I found that the texts frequently describe practical uses, especially for food, protection of water sources, and shelter. For example, in Genesis God declares: "I give you every seed-bearing plant on the face of the whole earth and every tree that has fruit with seed in it. They will be yours for food" (1:28–30). The Qur'an evokes a lovely representation of the bounty of trees: "Grieve not, for thy lord has placed a stream beneath thy feet; and shake towards thee the trunk of the palm tree, it will drop upon thee fresh dates fit to gather" (19:25–26). In his writing on Islamic environmental ethics, Mawil Y. Izzi Deen makes an even stronger statement on the importance of planting trees, saying that "even when all hope is lost, planting should continue for planting is good in itself. The planting of the palm shoot continues the process of development and will sustain life even if one does not anticipate any benefit from it." The prophet Muhammad said, "When doomsday comes, if someone has a palm shoot in his hands, he should plant it."

In my searches of the Old Testament, I found that these sacred writings contain many metaphors equating trees with bounty. The righteous are compared to a tree planted by a stream, always fruitful and with well-watered roots, whose foliage will never wither (Psalm 1; Jeremiah 17:8). The righteous are also said to thrive like a palm and to grow tall as a cedar in Lebanon, full of sap and richness (Psalm 92) or like a leafy olive tree (Psalm 51). The presence of grapes on people's vines or figs on their trees often denotes safety and prosperity. The prophet Micah envisioned a time of world peace, when everyone shall "sit under his vine and fig tree and none make them afraid" (Micah 4:4).

Trees also served the ancient Hebrews as geographical reference points: "Abram traveled through the land as far as the site of the great tree of Moreh at Shechem" (Genesis 12:5–7). Many passages describe decorations in the shape of trees on objects and structures: "On the walls all around the temple, in both the inner and outer rooms, he carved cherubim, palm trees and open flowers" (1 Kings 6:28–30); or "the face of a man toward the palm tree on one side and the face of a lion toward the palm tree on the other were carved all around the whole temple" (Ezekiel 41:19).

So essential and useful are trees that the Bible instructs Hebrew warriors to spare them: "When thou shalt besiege a city a long time, in making war against it to take it, thou shalt not destroy the trees thereof by forcing an axe against them: for thou mayest eat of them, and thou shalt not cut them down (for the tree of the field is man's life) to employ them in the siege" (Deuteronomy 20:19).

Sacred writings of Islam likewise express strict rules of combat, rules that similarly include prohibitions against destroying trees. Soon after the prophet Muhammad died, Abu Bakr, his trusted caliph and friend, guided the Muslim army to defend its lands. His final instructions to the troops is a code of conduct in war that remains unsurpassed to this day: "Do not be deserters, nor be guilty of disobedience. Do not kill an old man, a woman or a child. Do not injure date palms and do not cut down fruit trees. Do not slaughter any sheep or cows or camels except for food. You will encounter persons who spend their lives in monasteries. Leave them alone and do not molest them."

The spiritual or religious value we assign to particular objects, such as trees, is often indicated by their inclusion in rituals and celebrations. When I studied comparative religion in college, I learned that Judaism originated as the religion of a pastoral, agrarian people who lived their lives in close daily contact with the natural world and saw their God manifest in it. The legacy of these origins can still be seen today, in Jewish holy writings, philosophical and legal traditions, and prayers and celebrations. In 1916, my mother's parents fled the pogroms of Russia to start a new life in New York City. My mother grew up in Brooklyn, speaking Yiddish and strictly observing the Sabbath. As a child, I loved to imitate the guttural cadence of Yiddish, the language she spoke with her own mother—my Bubby—and to hear about the holiday celebrations of her childhood: lighting the Chanukah menorah, eating *hamentaschen* (sweet buns) on Purim, and hiding the matzoh for the Passover seder.

As a child who climbed trees, I thought that Tu B'Shvat, the New Year of the Trees, was the best of the Jewish holidays by far. It falls on the fifteenth day of the month of Shvat, the fifth month of the Jewish calendar—typically, around mid-January. Tu B'Shvat's beginnings were strictly secular. The Torah required farmers every year to give a tenth of all crops grown to the priests of the Holy Temple, and Tu B'Shvat marked the start of the new tax year: all the fruits that grew from that day on were factored into the following year's tithe. It came to be celebrated as a minor holiday during the Middle Ages, but in the 1600s the Kabbalists, a mystical sect of Judaism that seeks insights into divine nature, carried it a step further. Concerned with *tikkun olam*—the spiritual repair of the world—they regarded honoring trees as one way of improving their spiritual lives. The Kabbalists created a Tu B'Shvat seder, which involves drinking four glasses of wine and eating many different fruits while reciting verses. (Eating fruit is a way of expiating the first sin, when Adam and Eve ate the fruit of the Tree of Knowledge.) Over time, the holiday became a day for celebrating trees, and beyond that, a day for celebrating Jews' connections to nature. In Israel and other countries, the day is commemorated with tree-planting ceremonies, and people give money to plant trees. Through these actions, modern Israelis affirm a future for their children, who will grow up enjoying the trees' fruit, shade, and beauty. One Tu B'Shvat blessing reinforces a daily mindfulness of God's presence in the natural world: "Blessed are you, Lord our God, spirit of the universe, whose world lacks nothing needful and who has fashioned goodly creatures and lovely trees that enchant the heart."

SPIRITUAL TEACHINGS FROM TREES

At times, the sounds and activities that surround religious and spiritual activities—the crescendos of high mass, the morning calls of the muezzin, and the jubilant chords of gospel music—can obscure the association of spirituality with quiet. In many cultures, however, a mindful focus on the breath is an important spiritual practice—for without breathing, life ceases. The word *spirit* derives from the Latin word *spirare*, to breathe, as do the words *spirituality*, *inspire*, and *expire*. The Hebrew words for breath, *nesheema*, and for soul, *neshama*, likewise share a single root. Trees serve as quiet reminders to slow down and breathe, sit in silence, and meditate. They breathe too, though they have no lungs or gills. Day and night, plants—just like animals—respire, taking in oxygen and releasing carbon dioxide, a process that provides energy for metabolism, growth, and reproduction. During the daytime, trees also carry out photosynthesis, harvesting energy from sunlight and converting it into sugars, which animals can then use to meet their energy needs. During this process, which is the complement of respiration, carbon dioxide is absorbed by the plant, while oxygen is exuded through stomata, the tiny pores on the underside of their leaves. Because the amount of oxygen a tree creates far exceeds its own respiratory needs, trees continually replenish the oxygen supply for humans and other animals. Thus every tree, every plant, every leaf becomes a connector among living things.

Every Tree

Every tree, every growing thing as it
Grows says this truth: You harvest what

You sow. With life as short as a half-
Taken breath, don't plant anything but

Love.

—*Rumi*

Like breathing, silence accompanies the sacred. We hush our voices when we enter a house of worship. The most powerful moments in ceremonies involving birth, marriage, or death are silent—the instant of inbreath just before the first cry of a newborn, the moment preceding the exchange of marriage vows, the interval when heads are bowed in remembrance of the dead. When I walk in the forest, silence is the companion I most avidly seek. I find it when I watch the concentric circles of a plunked pebble expand noiselessly in an alpine lake. When I see an isolated elm on a busy urban street, I experience both physical and spiritual peace amid the tumult of my noisy everyday existence.

In my childhood modern dance lessons I learned about "still points"—moments in the dance that are choreographed to be utterly still: although the music continues, the movement stops. Like punctuation in a short story or line breaks in poetry, still points provide a striking counterpoint to the movement that goes on before and after. Lately, I have become more aware of the excessive speed with which I live, and have consciously tried to slow myself down. I find myself looking for still points in my days and years, stopping to pause amid the tasks and tugs of a busy family and professional life. Happily, there are silent, yet vivid, reminders all around me, including the trees outside my window, in the parking lot, even in the articles I read. These serve as the still points in my day.

I Go Among Trees and Sit Still

I go among trees and sit still.
All my stirring becomes quiet
around me like circles on water.
My tasks lie in their places
where I left them, asleep like cattle. . . .

Then what I am afraid of comes.
I live for a while in its sight.
What I fear in it leaves it,
and the fear of it leaves me.
It sings, and I hear its song. . . .

—*Wendell Berry*

One branch of Buddhism is Zen, a term that comes from the Sanskrit *dhyana*, meaning "meditation." There is no god to worship, no formal rites to observe, and no future abode to which the spirits of the dead are destined. A major piece of Zen practice is *zazen*, or sitting meditation, which uses the breath to steady the mind. The essential point is to strive to be aware of and let go of distractions such as thoughts, emotions, and images. Zen Buddhists revere silence. They pay attention to air entering and exiting their nostrils to initiate and maintain a meditative state. This practice—of stopping, calming, and concentrating—is critical to their spiritual development. Although I know of no Zen practice that specifically includes trees, the outer stillness and silence that trees embody evoke the inner quiet that practitioners emulate in their meditative state.

Another Eastern philosophy is the Tao (the "path" or the "way," also called the Dao), founded by the Chinese master Lao Tzu. The aim of Taoism is to perceive the world as it actually exists, to understand limitations, and to see our life's path with clarity and simplicity. One important concept is chi, the vital energy of the universe that exists in all things, both living and inanimate. Chi, with its cosmic origin, is elusive and mysterious. It is represented as an aura generated by a body. Often, people practice seeing auras using large trees because it is said that a tree's chi is strong. It apparently looks much like a candle flame, with tongues of light extending outward. Those seeking to see auras are directed to find a group of trees during the transition between night and day. Seekers should be relaxed, balanced, and centered. If they pay attention and soften their sight, they should see a "flaming off" of the tree's chi toward the heavens.

Conversation among Mountains

You ask why I live
in these green mountains

I smile
can't answer

I am completely at peace

a peach blossom
sails past
on the current

there are worlds
beyond this one

—*Li Po*

The speaker in Li Po's poem seems to have found both useful knowledge and some deep sense of peace by dwelling on his green mountain. When I read this piece, I imagine a small-statured, simply clad man with a wooden staff walking slowly back to his sanctuary, the dark green of deciduous tree crowns shading his path. I see a small smile on his face, not the kind that responds to a joke, but rather one that reflects the recent memory of flower petals moving downstream before him, teaching him a simple lesson about time. Hermann Hesse also suggested that we use trees to help us find the answer to spiritual questions: "For me, trees have always been the most penetrating preachers. Trees are sanctuaries. Whoever knows how to speak to them, whoever knows how to listen to them, can learn the truth. They do not preach learning and precepts, they preach, undeterred by particulars, the ancient law of life."

One thing that tree form and function teach us is that living things are connected to each other, often in hidden ways. Although we each have separate perceptions, consciousness, and sensations, no individual exists apart from the world. This fact is echoed in food webs, ecosystems, and global ecology. Plants, animals, fungi, microbes, bacteria, and humans are all linked. No action we take will be without an effect on another being. The fossil fuels we burn in the United States contribute to global warming, which is submerging a Pacific atoll tens of thousands of miles away.

Observations I have made on an obscure canopy-dwelling plant support this idea. In lowland forests of the Pacific Northwest, the licorice fern grows on the branches of mature bigleaf maples, embedding their succulent licorice-flavored rhizomes (rootlike structures) in thick mats of moss and associated arboreal soil and organic matter. The graceful green fern fronds proliferate during the winter, with individual stems popping up in the early autumn when the winter rains start and then dying back with the onset of the summer drought. Like other botanists, I had always assumed that these stems were independent individuals, each with a unique genetic configuration. But one day when I climbed a maple tree to make some routine measurements of moss growth, I stuck my fingers into the moss mats to trace the root line of a fern frond. I traced it to the base of an adjacent frond, which was attached to the same rhizome. That root led to another shoot attached to the same root, and on and on, from branch to branch, connecting fern to fern all over the tree. Although fern fronds appear to be single individuals, they are actually interconnected physically and genetically—as am I, to my family, friends, colleagues, students, and, in a way, to trees.

A second lesson from trees is that we must be aware of things that are hidden. The belowground world represents the elements that we hide from ourselves and others—our troubles, ill health, addictions, and weaknesses. At times, we also conceal our greatest hopes and desires, our deepest personal truths, our untapped abilities, and those things that are most sacred to our spirits. The word *truth* is derived from the Old English *treowth*, meaning "fidelity," which was derived from the Sanskrit *daruna*, meaning "hard," and *dāru*, meaning "wood." We must recognize that these hidden parts are important parts of ourselves and not something to discount.

Late in his life, the poet Rainer Maria Rilke maintained a correspondence with a young poet named Franz Xaver Kappus. In some of his letters, Franz describes feelings of emptiness and loneliness, which he confesses he tends to bury. In a letter from Rome dated December 23, 1903, Rilke suggests that his young friend try to learn from his feelings of emptiness, and so, perhaps, come to cherish them: "What is necessary, after all, is only this: solitude, vast inner solitude." The solitary individual, he assures Kappus, has no cause for anxiety or sadness. "The nights are still there, and the winds that move through the trees and across many lands; everything in the world of Things and animals is still filled with happening, which you can take part in."

Forests perfectly embody what Rilke calls "filled with happening," the dynamism of life—though that very dynamism is something we humans often find difficult to accept. Whenever I revisit a favorite forest trail, I regreet the same trunk, the same stream, the same deep smells I noted the month before, and that gives me a sense of stability. However, trees change—a lot, and sometimes quickly. Although seasonal patterns are most pronounced in deciduous forests, even conifer trees, which seem not to lose their foliage, reinforce the law of life that everything must change. In the Pacific Northwest, where I take hikes and trail runs, it is the mosses that grow on trees that provide the greatest evidence of change. In the dry summer months, the mosses go into a kind of dormancy, becoming drier and drabber as the season progresses. Winter, when the rains bring moisture and nutrients, is when the mosses explode silently with growth, putting out new fronds and spore-bearing stalks in brilliant abundance.

Seeking a spiritual life consists of finding and adhering to a moral code that provides a personal and communal sense of right and wrong. Humans draw those precepts from a wide variety of sources, creating icons or symbols to represent our sense of these things. Surprisingly often, these symbols relate to trees or patches of forest, often ones encountered in childhood. I have my own memories of special trees, which have held me through difficult times. The maple tree that stood outside my childhood home and tapped companionably on my bedroom window kept me company on scary, windy nights, assuring me there was something strong and solid out there in the dark; it kept me from feeling alone and connected me to the world of nature and life that existed outside my little room. Even if you don't have such a tree of your own yet, David Wagoner's poem "Lost" assures you that those trees are there, looking for you:

> Stand still. The trees ahead and bushes beside you
> Are not lost. Wherever you are is called Here,
> And you must treat it as a powerful stranger,
> Must ask permission to know it and be known.
> The forest breathes. Listen. It answers,
> I have made this place around you.
> If you leave it, you may come back again, saying Here.
> No two trees are the same to Raven.
> No two branches are the same to Wren.
> If what a tree or bush does is lost on you,
> You are surely lost. Stand still. The forest knows
> Where you are. You must let it find you.

Connections between trees and the spirit are as universal as tree form itself. Although some of the bitterest battles have been fought on religious grounds, the spirituality of trees may transcend differences that separate the world's faith traditions. The spiritual teachings of trees are universal: we should strive to connect the mundane with the heavenly, produce things that are useful to others, be rooted in our home place, accept the inevitable changes of life, live mindfully, be joyful. Opening up to something as simple and pleasurable as climbing a tree—or sitting silently beneath it—can make humans feel at home with the world, and with themselves.

Chapter Nine

MINDFULNESS

> Let me desire and wish well the life
> these trees may live when I
> no longer rise in the mornings
> to be pleased by the green of them
> shining, and their shadows on the ground,
> and the sound of the wind in them.
>
> —*Wendell Berry, "Planting Trees"*

I lie dying.

I can see it in my mind.

I am tied to a broad branch of my favorite fig tree high above the tropical rainforest floor at my study site in Costa Rica. My husband, Jack, gently tightens the straps a final time, kisses me again, and we say good-bye, recalling all the years and all the love between us. He sets his rappelling gear in place and slides down the rope to the forest floor. I hear the rustle of the leaves beneath his footsteps on the trail—pausing once, twice, then leaving me. When we were married in a giant ceiba tree in 1983, I asked Jack to do exactly this when I got old and death approached me. Now, after his descent, I hear nothing but the sound of one thousand thousand leaves speaking to me in the language that I have tried to learn in my life of climbing trees. In a day or so, I will slip away from life, and the elements that make up my body will join the nutrients of the branches, the moss, the canopy earthworms, the canopy birds, and onward into the forest food web. To me, this is the best possible death, to finally become a tree in the way Robert Morgan described in his poem "Translation":

> Where trees grow thick and tall
> In the original woods
> The older ones are not
> Allowed to fall but break
> And lean into the arms
> Of neighbors, shedding bark
> And limbs and bleaching silver
> And gradually sinking piece
> By piece into the bank
> Of rotting leaves and logs
> To be absorbed by next
> Of kin and feeding roots
> Of soaring youth, to fade
> Invisibly into
> The shady floor in their
> Translation to the future.

Perhaps it is a professional mistake to admit to such intense feeling for my subject of study. But my life and work—and this book—have been all about the intertwining of people and trees. By studying the ecology of trees, and perhaps even more by exploring their myriad links to the cultural, aesthetic, and social aspects of humans, I have become more mindful of my responsibilities to them, whether in my family, in my community, or on my planet. This mindfulness—being aware and compassionate—occupies the apex of the modified Maslow pyramid I discussed in the introduction. And it is with mindfulness that I choose to close this book.

THE IMPORTANCE OF TREES

> On the last day of the world
> I would want to plant a tree
> What for
> Not for the fruit

The tree that bears the fruit
Is not the one that was planted
I want the tree that stands
In the earth for the first time
With the sun already
Going down
And the water
Touching its roots
In the earth full of the dead
And the clouds passing
One by one
Over its leaves

—*W. S. Merwin, "Place"*

One way to gauge the importance of something in our lives is to recognize the impact of its absence. For example, if you break your seemingly insignificant pinkie toe, you soon learn, as you hobble about your daily rounds, how much you normally use it. To assess the importance of trees to our essence as humans, we can consider what the world might be without them. The novelist Cormac McCarthy presented a powerful treatment of this theme in his Pulitzer Prize–winning novel *The Road.* In this dark vision of the future, an unspecified environmental disaster has rendered the world nearly uninhabitable. The few remaining humans scuttle across a barren landscape, scavenging for scant food and avoiding brutal, cannibalistic bands of survivors. The story revolves around a man and his son who are walking south along a ruined road to escape the cold and dark of the nuclear winter that has gripped the planet. To me, the horror of McCarthy's vision is reinforced by his frequent descriptions of dead trees and dying forests: "charred and limbless trunks of trees stretching away on every side"; "barren ridgeline trees raw and black in the rain"; "the trees in the orchard in their ordered rows gnarled and black and the fallen limbs thick on the ground"; "thin black trees burning on the slopes like stands of heathen candles"; burnt forests for miles along the slopes and snow sooner than he would have thought; a distant stand of trees and sky faded stark and black against the last of the visible world.

For the two protagonists, what little relief they find often comes from trees. Firewood, a gift of the dead trees to the surviving humans, is a life- and spirit-sustaining commodity: "He scuffled together a pile of the bone-colored wood that lay along the shore and got a fire going and they sat in the dunes with the tarp over them and watched the cold rain coming in from the north." The only other comfort the man has is his memories from the past. One in particular concerns the annual trip he took as a child with his uncle to a lake to cut and gather firewood, "the shore lined with birches that stood bone pale against the dark of the evergreens beyond. . . . This was the perfect day of his childhood. This the day to shape the days upon."

A second way to understand the importance of something in our lives is to recognize that the prevention of contact with it—whether deserved or undeserved—contributes to a sense of punishment. This has occurred over and over in history when people were forcibly removed from their natural surroundings. During the Nazi occupation of Holland, for example, Anne Frank, whose diary made vivid the life of a family forced into hiding for twenty-five months, described the old chestnut tree from the one window in their cramped quarters that was not blacked out. On February 23, 1944, she wrote: "Nearly every morning I go to the attic to blow the stuffy air out of my lungs. From my favorite spot on the floor I look up at the blue sky and the bare chestnut tree, on whose branches little raindrops shine, appearing like silver, and at the seagulls and other birds as they glide on the wind. . . . As long as this exists, and I live to see it, this sunshine, the cloudless skies, while this lasts I cannot be unhappy." Despite the darkness of her final months, in her diary Anne expressed a strong faith in the world's essential goodness—an attitude evoked also in this poem by Wendell Berry:

Woods

I part the out thrusting branches
And come in beneath
The blessed and blessing trees.
Though I am silent

There is singing around me.
Though I am dark
There is vision around me.
Though I am heavy
There is flight around me.

The connection between Anne and her now 150-year-old chestnut tree has resonated with the readers of her diary over the decades since her death in 1944. However, the tree itself has not fared well. In 1993, soil analysis showed that leakage from a nearby domestic fuel tank had endangered its root system, and more recently the tree has been attacked by a fungus that has further undermined its stability. In addition, the horse-chestnut leaf miner, a small but voracious insect, has caused premature leaf fall, weakening the tree even more. In March 2007, the Amsterdam City Council declared that the tree must be cut down. This prompted extended protests by the Netherlands Tree Institute and worldwide pressure from people who see the tree as a symbol of hope during the darkest of times. Despite massive efforts to preserve the tree, including soil sanitation and extensive trimming of the crown, Anne's tree appears to be unsavable. However, the tree won several legal reprieves in 2007, and in November Judge Jurjen Bade ruled that the tree posed no immediate danger and called for alternative measures to be explored. He said that felling the tree should be a "last resort" and told the council to meet with conservationists to find a solution in the following months.

Prisons are another place where contact with nature is withheld, consciously or not, as punishment. This has always puzzled me, given plants' association with regeneration and renewal. If we could encourage not only prisoners but also their jailers to value the healing qualities of nature, perhaps the strikingly high recidivism rate—nearly 40 percent in my home state of Washington—would decline. Some prisons do have programs in which prisoners cultivate gardens that supply the prisons with fresh vegetables, which is constructive and healthy. However, interacting with plants in ways that go beyond gardening can secure other intellectual and emotional benefits.

Several years ago, I developed a "Moss-in-Prisons" project that brought prisoners together with living, growing things that needed their care. I chose mosses for two reasons. First, I realized that it was unrealistic to introduce trees within prison walls. But even more important, wild mosses from old-growth forests were in need of help. In recent years, a thriving moss-collecting industry has developed because of demand from florists, who use moss for flower arrangements and for packing bulbs. As of 2005, moss harvesting had an annual economic value of between $6 million and $260 million. The wide range of the estimates illustrates how little is known abut the moss harvest trade.

Because these plants fill important ecosystem roles and are very slow to regenerate, ecologists were already voicing concern over expanded harvesting of this "secondary forest product." Epiphytic mosses capture and retain nutrients that occur in dissolved form in rainfall and mist. When they are alive, the mosses are held in the living biomass, but when they die and decompose, those elements enter the soil and are taken up by terrestrially rooted plants, which are then eaten by herbivores such as deer and insects, and subsequently cycle through the ecosystem. Thus, these mosses are a critical entry point for scarce elements that contribute to the nutrient cycles of the whole forest. In addition, they provide habitat for arboreal invertebrates such as earthworms and beetle larvae, and supply food and nesting spots for arboreal mammals and birds such as the endangered marbled murrelet. Recent studies have shown that moss communities take decades to grow after disturbance, far longer than would allow for sustainable harvest at present removal rates. No protocols exist for growing mosses commercially. To learn how best to cultivate usable moss in sufficient quantities to offset harvest from the wild, I needed help from people who have long periods of time available to observe and measure the growing mosses; access to space where flats of plants can be laid out; and most important, fresh eyes and minds to spot innovative solutions. These qualities, I thought, might be shared by many people in prison.

The biology of mosses also makes them suitable for novice botanists, because mosses possess "poikilohydric" foliage, which means their thin foliage wets and dries rapidly, allowing them to survive drying without damage and to resume growth quickly after rewetting. Some mosses that have lain in herbarium drawers for over one hundred years have been revived by simply applying a little water and bringing them into the light. They therefore tend to be resistant to under- or overwatering, increasing the probability of their survival in a prison setting. Learning how to grow mosses in captivity would also provide an opportunity for people with no access to nature and little opportunity to use their intellects to learn about scientific research and plant conservation.

After scouting prisons in my region, I found the Cedar Creek Correctional Center in Littlerock, Washington. The superintendent, Dan Pacholke, is a second-generation prison official; he is a soft-spoken man whom I would expect to encounter behind the reference desk of a library rather than directing an operation that oversees four hundred convicts. From the beginning, he was enthusiastic about the project and facilitated all aspects of it, forging pathways through the Department of Corrections administration. Our moss-growing team included a warden, two of my students, an adult volunteer, and six inmates who rotated in as their fellow prisoners' sentences were completed.

We started with a straightforward goal: to explore new ways to "farm" moss that would provide an economically viable way to alleviate collecting pressure on the ancient forests of the Pacific Northwest. A first critical step was to find out which of the hundreds of species of moss were most amenable to horticulture. We also needed to learn the most basic aspects of caring for them: How much water and nutrients do different kinds of mosses need? Should solutions be delivered as droplets or as mist? If I could answer these questions—with the collaboration of my incarcerated partners, who could observe and record the growth rates of our experimental plants—then we would be closer to solving a hitherto intractable ecological and conservation problem.

My students collected a variety of moss samples with a permit from the Olympic National Forest and brought the fragrant sacks of plants to the spare prison yard, where we were surrounded by concrete walls and barbed wire. The prisoners quickly built a small greenhouse with recycled lumber and we were set to go. Each inmate had a notebook and pencil to record his observations. The prisoners quickly learned to identify the four moss species we had collected, using their scientific names (only a few mosses have common names). They devised their own ways to grow them—hanging clumps of moss in mesh bags, for example—and contrived ways to deliver water with medical tubing and hardware clamps. They also learned how (and why) to retrieve randomized subsets of mosses to air-dry for growth measurements. Each month, my students and I took a set of moss samples to my lab at the Evergreen State College to dry and weigh, for a comparison sample.

The results of the project were dramatic in many ways. After eighteen months, we all shared the excitement of knowing which mosses grow fastest (*Eurhynchium oreganum* and *Rhytidiadelphus loreus*) and which watering treatment is most effective (misting proved better than droplets). We have since been working with two online nature gift companies to market "sustainably grown moss pots" using the plants the prisoners have propagated. "NatureLink" hangtags accompany each pot to inform the person buying it of the ecological importance of mosses and the need to grow them sustainably. When we finalize negotiations with these companies, we hope that some of the prisoners will pursue this alternative line of employment after their release.

Other rewards came from this project, ones I had not foreseen. Nearly all of the inmates (whose names I have changed here) gave me insights into the ways that working with plants had affected them. One of them, Wayne Hunter, joined the horticulture program at the local community college after his release, with a career goal of opening his own plant nursery. "I don't want to just mow lawns and trim hedges," he said firmly. "I want to grow real plants." Another, José Juarez, told me he had taken an extra mesh bag of moss from the greenhouse and placed it in the drawer of his bedside nightstand. Each morning, he said, he opened the drawer to see if the moss was still alive. "And though it's been shut up in a dark place for so long, it's still alive and growing this morning" he said, grinning. And then, more quietly, "Like me."

TREES AS TEACHERS OF MINDFULNESS

Soak up the sun
Affirm life's magic
Be graceful in the wind
Stand tall after a storm
Feel refreshed after it rains
Grow strong without notice
Be prepared for each season
Provide shelter to strangers
Hang tough through a cold spell
Emerge renewed at the first sign of spring
Stay deeply rooted while reaching for the sky
Be still long enough to hear your own leaves rustling.

—*Karen Shragg, "Think Like a Tree"*

The ex-prisoners, Wayne and José, learned lessons about mindfulness under difficult circumstances from the mosses they cared for. Many other people have written about similar lessons. For example, a millennium ago Saint Bernard of Clairvaux, in Epistle 106, wrote, "You will find something more in woods than in books. Trees and stones will teach you that which you can never learn from masters." Henry David Thoreau, in *Walden*, averred that nature teaches us the basics of life and the ways to live it: "I went to the woods because I wished to live deliberately, to front only the essential facts of life, and see if I could not learn what it had to teach, and not, when I came to die, discover that I had not lived."

One mindfulness lesson has stayed in my memory since I read Antoine de Saint-Exupéry's *The Little Prince* in my seventh-grade French class. In the book, an airplane pilot narrates the story of the travels of the mysterious and insightful Little Prince, a little boy who is more than just a little boy, whom the pilot encounters when his airplane is downed in the desert. In the first chapter, the Little Prince explains that baobab trees have taught him about the need to pay attention to things that can become harmful if we are not mindful of them. After leaving his own little home planet, the Little Prince worried about how vulnerable his planet and his beloved rose were to the invasion of baobab seedlings. "Indeed, as I learned," narrates the pilot,

> there were on the planet where the little prince lived—as on all planets—good plants and bad plants. In consequence, there were good seeds from good plants, and bad seeds from bad plants. But seeds are invisible. They sleep deep in the heart of the earth's darkness, until some one among them is seized with the desire to awaken. Then this little seed will stretch itself and begin—timidly at first—to push a charming little sprig inoffensively upward toward the sun. If it is only a sprout of radish or the sprig of a rose-bush, one would let it grow wherever it might wish. But when it is a bad plant, one must destroy it as soon as possible, the very first instant that one recognizes it.
>
> Now there were some terrible seeds on the planet that was the home of the little prince; and these were the seeds of the baobab. The soil of that planet was infested with them. A baobab is something you will never, never be able to get rid of if you attend to it too late. It spreads over the entire planet. It bores clear through it with its roots. And if the planet is too small, and the baobabs are too many, they split it in pieces . . .
>
> "It is a question of discipline," the little prince said to me later on. "When you've finished your own toilet in the morning, then it is time to attend to the toilet of your planet, just so, with the greatest care. You must see to it that you pull up regularly all the baobabs, at the very first moment when they can be distinguished from the rosebushes which they resemble so closely in their earliest youth. It is very tedious work," the little prince added, "but very easy."

TAKING MINDFUL ACTION

> If you want to build a ship, then don't drum up men to gather wood, give orders, and divide the work. Rather, teach them to yearn for the far and endless sea.
>
> —*Antoine de Saint-Exupéry,* The Wisdom of the Sands

From the *Little Prince* passage I learned how important it is to be mindful, not of baobab seedlings per se (not on my planet, anyway), but of all things that might start out as small and harmless actions but have the potential to grow large and horrible—consumerism, violence, racism, addiction, fascism—all of which must be dealt with before they overtake our personal and societal landscapes. On our own small home planet, two of the most severe threats to trees and therefore to ourselves are consumerism and population growth, which are fueling, among other things, deforestation, forest fragmentation, overharvesting, and climate change. Cadres of environmental scientists, sociologists, geographers, and policymakers are assessing and communicating the declining state of our planet—its trees and other aspects of our natural environment. Many remedies have been proposed, from changing standard forest practices to shifting to alternative fuel sources.

To be honest, I am sometimes at a loss to present succinct and useful recommendations for weeding out these baobab seedlings. As a forest ecologist, I read the scientific facts and figures and can understand and discuss in an academic context the problems that trees face. But I also witness the growing emotional disconnection and distance between people and nature. All of the peer-reviewed academic graphs and tables I can muster only rarely bridge the gap to make much of a difference in the way people act. It seems we need a combination of evidence from the intellect and impetus from the heart to shift the ways humans interact with trees. It also seems that individuals need neither preaching nor forceful directives to make changes, if they themselves become convinced of the need to maintain trees, forests, and nature in their lives. I believe the answers must begin with mindfulness—the topmost part of Maslow's modified pyramid—on the part of each individual.

One way to begin to formulate a plan for action to protect and wisely use trees is to look at the actions of individuals. There is a long history of people who have taken mindful action on behalf of trees. For example, in 1670 the great Maratha emperor Shivaji of India posted a sign at the entrance to one of his forests giving instructions on the permitted harvesting of valuable trees such as teak. He declared these to be "reserved against felling for they are slow-growing. Any act of irrational cutting of trees is a matter of grief. Indiscriminate cutting of trees is calamitous. It is the citizen's duty to guard our forest wealth." The emperor's foresight spared some of the most beautiful and extensive natural teak forests of central India, which are still intact and continue to support populations of Bengal tigers, leopards, and gaur (the wild Indian bison).

In Africa in the present day, we encounter another powerful advocate for conservation. Wangari Maathai, a Kenyan activist, observed how forest loss has led to desertification in Africa and now threatens many other regions of the world. She also recognized that trees help farmers by improving the water cycle, protecting soil, and enhancing wildlife habitat. Her work began when she planted nine tree seeds in the yard of her own house. In ensuing years, she and her co-workers have persuaded village women across Africa to fight deforestation. Rural women, whose traditional duties include collecting firewood, quickly understood that planting trees would help farmers, ensure a supply of wood, and strengthen their own place in society.

In 1977, Maathai founded the Green Belt Movement, whose mission was to empower women, confront corrupt officials, and plant thirty million trees. The organization has given jobs to ten thousand women, who grow and sell seedlings. In 2004, Maathai became the first African woman to receive the Nobel Peace Prize. By choosing to honor her and her work, the Nobel Committee elevated all people who work to protect trees and their environments. Maathai has inspired international groups to launch similar tree-planting initiatives in over forty other countries, and she has been awarded many international environmentalist and humanist prizes and honorary degrees for her work. She stated, "Today we are faced with a challenge that calls for a shift in our thinking, so that humanity stops threatening its life-support system. We are called to assist the Earth—to heal her wounds, and in the process heal our own."

Another woman who has taken unique action for trees is Julia Butterfly Hill. Her two-year sit atop a redwood in Humboldt County, California, generated international media attention and raised public awareness of the importance of old-growth trees. On December 10, 1997, the

29. Kenyan Wangari Maathai received the Nobel Peace Prize in 2004 for her pioneering work inspiring women—especially poor women—to plant seedlings in over forty countries around the world. Reproduced by permission of the Green Belt Movement.

twenty-three-year-old climbed into a thousand-year-old, 180-foot-tall coast redwood tree in order to prevent the Pacific Lumber Company from logging the tree, which she named Luna. She descended 738 days later, on December 18, 1999, when the Pacific Lumber Company agreed to preserve Luna and all trees within a three-acre buffer zone. Earlier that year, Hill and other activists founded the Circle of Life Foundation, educating people about the interconnection and interdependence of all life. Her nonviolent actions in defense of the forest entitled her to be the youngest person inducted into the Ecology Hall of Fame.

Julia was not the first to make tree sitting an effective conservation technique, however. One of the earliest expressions of tree activism was Almstriden (Battle for the Elms), staged in Stockholm, Sweden, in May 1971. People sat in hammocks and chained themselves to trees to prevent the city officials from cutting down a group of elms to construct a subway station. This protest led to the founding in 1995 of Ecopark, a national park that protects nature and culture in the middle of Stockholm. It also marked the beginning of a European grassroots campaign to protect nature from urban development. In France, for example, Grenoble's Paul Mistral Park was the focus of a 2004 demonstration by thousands of "eco-citizens" protesting the planned destruction of more than three hundred trees to make way for a giant soccer stadium. Ten tree platforms were linked by hanging bridges, and the people of Grenoble showed their support for the arboreal village by bringing supplies of baguettes, cheese, and bottles of champagne. The protest collected 15,000 signatures calling for a referendum to preserve the land for a park. This and other grassroots campaigns demonstrate that fairly small numbers of individuals can garner the attention needed to bring about positive change.

Another heroine of tree activism has made quieter contributions to forest conservation. Ingrid Gordon, a long-time Greenpeace activist and rock climber, knew how difficult it was for grassroots organizations to afford high-quality equipment. In 2001 she founded Gear for Good, which connects activists and researchers on a limited budget with outdoor gear companies. Recipients have included tree conservation groups such as the Ruckus Society, Great Old Broads for Wilderness, the Alaska Coalition, the Northwest Ecosystem Alliance, and the International Canopy Network. This remarkable woman makes her donations without fanfare, and reminds me that the work of stewardship draws together people of diverse experience, skills, and personalities, then branches out into actions as quiet and significant as trees themselves.

As a canopy researcher who has benefited from Ingrid's work, having been outfitted by Gear for Good to climb into trees and collect ecological data, I sometimes worry that my years of research are of minimal or only indirect use to protect trees. Counting hummingbird visits to flowers in a remote cloud forest can seem hopelessly obscure when juxtaposed with photographs of massive deforestation. And yet forest canopy researchers have begun to contribute both to increasing mindfulness and to active protection of trees at the global level. In 1994, my then–graduate student Joel Clement and I started a small but dedicated nonprofit organization, the International Canopy Network (ICAN). Since then, it has grown into an active and comradely society whose mission is to enhance communication among researchers, educators, arborists, and conservationists concerned with forest canopies. We produce a quarterly newsletter called "What's Up?," maintain a website, manage an international electronic mailing list, inform the media about canopy matters, and promote awareness through activities such as Washington State Forest Canopy Week (which, held every July 17–24, we successfully lobbied for in the legislature). Most important, we maintain a growing database—now over seven thousand entries strong—of scientific and popular articles, books, and films that deal in some way with the forest canopy. The outreach efforts of ICAN reach a wide range of people, from preschoolers to senior citizens.

Important things are being done at the global scale as well. In 2001, policymaker Andrew Mitchell and an international group of canopy researchers founded the Global Canopy Programme (GCP). Based in Oxford, England, this alliance links studies of worldwide forest canopies and questions relating to biodiversity, climate change, and poverty alleviation. In 2005, the group represented the forest canopy research and conservation community at the United Nations Climate Change Convention conference in Nairobi, which was attended by six thousand delegates from 180 countries. Andrew argued that forest canopies offer significant ecosystem services to humanity, including carbon storage (about two hundred tons of carbon per hectare), rainfall generation, on which agricultural security and energy security depend, water storage (three trillion tons in the Amazon alone), cooling to buffer regional weather conditions, and biodiversity (an estimated 40 percent of the global total)—and that these services may become tradable between forest-rich countries and resource-poor countries in the future.

In 2007, the GCP organized canopy scientists, economists, and policymakers to create the Forests Now Declaration, a mechanism to raise awareness of the importance of forests and forest canopies worldwide. The declaration, simply put, is a call to recognize "forests in the fight against climate change." The document presents the case that human-induced change is real and that tropical deforestation contributes nearly one-quarter of global carbon emissions, second only to fossil fuel burning. The signers state that tropical forests offer one of the largest opportunities for cost-effective action and must now be treated with urgency. They point out that tropical deforestation and forest degradation are driven by external demands—for timber, beef, soya, and biofuels—further raising the stakes of global warming. However, residents of tropical countries are too often excluded from carbon markets that could provide needed alternative strategies because of the complexity of carbon market rules and because of short-term economic incentives to harvest their forests and enhance their trade balance with other countries. The Forests Now Declaration calls on governments to carry out three actions: (1) encourage mechanisms to recognize forest ecosystem services, (2) help developing nations to fully participate in carbon markets, encouraging trade with countries that have capital to invest in carbon storage, and (3) provide inducements for sustainable use of ecosystems, including removing incentives that encourage forest destruction. The measure has been endorsed by representatives of over fifty-five scientific and conservation groups, including the Association for Tropical Biology and Conservation, Conservation International, the International Union of Concerned Scientists, and the Sumatran Orangutan Society. Among the individual luminaries who have put their signatures to the document are Wangari Maathai and Jane Goodall.

The stories of all of these individuals—Shivaji, Wangari Maathi, Julia Butterfly Hill, Ingrid Gordon, Andrew Mitchell, and Joel and me—illustrate my belief that when humans become sufficiently aware of the value of something, whether it be a single tree, a forest fragment, or the whole biosphere, they will find ways to protect it. Although many of these stories highlight big actions, when it comes to mindful activity, size does not matter. Sometimes, in fact, you have to start small. What is critical is to feel the commitment within yourself.

If we want to participate in maintaining the health of our intimate connections to trees, we must take action. Over the past few years, our newspapers, magazines, and billboards have been filled with advice and slogans to motivate and direct our efforts to change the way we consume, recreate, and transport ourselves to produce a smaller ecological footprint in our communities and on our earth. *Recycle! Buy shade-grown coffee! Ride a bike! Boycott furniture made of tropical hardwoods!* Nearly all of us have heard these exhortations, and many of us heed them.

One of the most direct conservation activities we can engage in is planting trees, a mindful act that requires only two hands, a shovel, and a seed or seedling. The National Arbor Day Foundation, established by J. Sterling Morton in 1872 "to inspire people to plant, nurture, and celebrate trees," promotes tree planting and tree awareness with remarkable enthusiasm. The organization distributes posters, installs tree memorials, organizes corporate tree-planting programs, and gives advice on how to plant the ten free trees that come with joining their group. Traditionally, Arbor Day is a day to plant trees in a public arena. I vividly remember the event in my second-grade class, during which we paraded around the schoolyard holding twigs in our hands and then witnessed Mr. Newcomb, our principal, shovel the rain-soaked soil of spring to plant a young cherry tree north of the playground, looking odd but wonderful with his suit and necktie and soil-specked shoes. The tree he planted stands there to this day, offering its flowers and shade to all who visit it.

Another way to direct mindfulness is to seek and develop creative collaborations with partners, especially those who might be outside your daily spheres. Many collaborative examples exist, but to me, one of the most intriguing and potentially effective conservation campaigns involves the coming together of environmentalists with conservative evangelical Christians, a seemingly odd couple given their historical differences. But as environmental losses and global climate change grow more pressing, the calls of environmentalists have become impossible to ignore, regardless of religion or politics. In February 2006, eighty-six evangelical leaders signed the Evangelical Climate Initiative, a landmark proposal that, in addition to urging believers to join the fight against climate change, calls for federal legislation to reduce carbon dioxide emissions. Among the signers of the statement are the presidents of thirty-nine evangelical colleges, leaders of aid groups and churches, and pastors of megachurches, including televangelist Rick Warren. Despite opposition from some church leaders, many evangelical churches are preaching a gospel of "creation care," an ethic inspired by scriptures wherein God gave people dominion over the earth, and with it an obligation to carry out good stewardship of the land, air, and water. This aligns evangelical goals with the actions of many grassroots environmental groups: cleaning up streams, planting trees, advocating for clean energy and against overconsumption and materialism.

However, caveats accompany the seemingly simple and goof-free practice of a planting a tree, advocated by so many recognized conservation groups. The critical message is to plant the right tree for the right place rather than sticking any old stick in the ground. It must grow well with minimal maintenance, as well as having a natural size limit suited to the space in which it has been planted. Because of the long lifetimes of trees relative to our own, it is easy to enthusiastically plant a tree that will either outgrow its surroundings or require topping or premature removal. I recently invited Micki McNaughton, an urban forester for the City of Olympia, to speak to my forest ecology class. She is a small, energetic woman whose enthusiasm for trees seems boundless. At one point in her lecture, she referred to a set of maple trees that line Legion Avenue, between East Bayview and Plum avenues, in downtown Olympia. All of the students immediately knew which trees she meant: a colonnade of imposing, broad-crowned trees that are hard to miss. Her face became stern and sad at the same time. "Terrible!" she snapped. "Those trees were topped decades ago because they were growing too high for the telephone wires and the houses on the street. Their tops were slashed, and now, each and every one of those trees is rotten at its core; their supporting heartwood has all but disappeared. I predict"—and here her sadness deepened—"that we will have to take each and every one of them down in less than a decade."

Later in her lecture, Micki emphasized that she had no wish to discourage us from planting trees. Indeed, tree planting by citizens can be a positive and powerful tool to help the environment and people. But she wisely advised my students and me to do so with the guidance of people like city or county arborists, cooperative extension managers from state universities, or certified arborists who have experience choosing the right tree, the right place, and, most important, the right maintenance for the very long time scale in which trees dwell.

She concluded with another example of misguided tree planting. In

1876, Ellwood Cooper planted fifty thousand eucalyptus seedlings, or "eucs," on his ranch near Santa Barbara, California. Thirty years later, they were 175 feet tall, an unprecedented growth rate for plants in those dry hills. By the turn of the twentieth century, groves of these trees were prevalent in California's coastal and central landscapes. Although Australia produced excellent lumber from its old-growth native eucalyptus forests, in America the fast-growing pioneer trees grew too quickly to yield useful lumber, as quick growth entails a critical reduction in the structural strength of the wood. As they continued to spread, their water-seeking roots blocked drains, damaged pavement, and fueled wildfires. The incendiary sap, gigantic volumes of leaf litter, and long shreds of hanging bark of these "gasoline trees," as firefighters call them, evolved millions of years ago to promote the fires that open their seedpods. But these qualities now endanger valuable urban and suburban property, as well as the native habitat around the trees. The litter layer can accumulate to four feet thick because the decomposer communities—microbes and insects—that in the trees' native Australia break it into soil do not occur here.

The costs of planting this unsuitable nonnative tree species have been high, especially when we add the damage they do to native trees and the organisms that depend on them. One of the habitats endangered by eucalyptus groves is coastal scrub, which supports such endemic plants as coast live oak, California bay laurel, coyote bush, wax myrtle, California sagebrush, mule's ear, and native bunchgrasses. The scrub has an understated beauty and value to wildlife that can be easily overlooked by a public more concerned with the size and stature of trees in a landscape. Although birds do frequent eucalyptus groves, ornithologists at the Point Reyes Bird Observatory have found that species diversity is 70 percent less than in stands of native trees. To deal with the sticky gum produced by eucalyptus flowers, native Australian honeyeaters and leaf gleaners have evolved long bills, but North American leaf gleaners such as kinglets and vireos have not, so the gum clogs their faces and bills, sometimes with fatal effects. The pluses and minuses of planting eucs have been as incendiary as their leaves and flowers. Some people love these trees for their ability to populate treeless areas, for their rich smell, their medicinal value, and, well, for their treeness in a dry landscape. But the consensus is leaning increasingly toward the ecologically correct decision to cease planting eucalypts in favor of native tree species. Although some people persist in planting this ecologically misplaced tree species, education programs are having an impact, and the tree is becoming the arboreal equivalent of a persona non grata in southern California landscapes.

Even when we consider the planting of native trees, we must keep in mind that many fare poorly in the inhospitable conditions of urban environments, even though they grew on the original terrain naturally. The archetypal urban conditions of compacted and dry soil, atmospheric pollution, and temperatures well above those in surrounding nonurban areas can easily cause failure, even of native trees. It is important to consider the actual environment in which the tree will spend its life.

BRANCHING OUT

There is another way for mindfulness to emerge and create action, an approach I am just beginning to explore in a formal way. I establish situations in which I can closely collaborate with artists so that I might literally see with new eyes and communicate with a chorus of new voices. In 2006, I had a remarkable opportunity to bring together two realms of my life that I had thought would be forever separate: forest canopy ecology and modern dance. For the past three decades, since deciding to leave dance in favor of field biology, I have visited that world only for an occasional class or to watch a professional performance. Then I received a telephone call from Jodi Lomask, the director and choreographer for a San Francisco–based modern dance troupe called Capacitor. She was interested in creating a dance about symbiosis and the tropical rainforest, and invited me to join her troupe for an intensive workshop in a Central American rainforest to develop it. My job was to provide scientific information about the ecological interactions they wished to incorporate into their movements.

A few months later, Jodi, her seven dancers, a filmmaker, and I found ourselves in Monteverde, Costa Rica, my long-term study site. We spent a marvelous ten days exploring and interpreting the flora and fauna through the medium of dance. I taught them how to climb, and we ascended the huge strangler fig I call Figuerola. The limbs of this cloud forest giant stretch horizontally in all directions and are covered with a riot of moss, ferns, bromeliads, shrubs, and orchids. Aloft, the dancers took it all in, though not with a measuring tape, clinometer, and Rite-in-the-Rain notebook, as I and my students would have done; rather, they recorded what they experienced with unfurling fingers, curved arms, and lifted eyebrows. After several days in the cloud forest, they peeled off their clothes to merge more fully with the forest textures. I was filled with wonder as I watched human limbs parallel tree limbs, human trunks intertwining with twisted tree trunks, and fingers waving like the gentle twigs of the understory trees.

At the end of our visit, the troupe invited the members of the Monteverde community to a preliminary performance. Although many in the audience had never seen modern dance before, it was clear that they "got it." The Costa Ricans were excited to see "their" cloud forest animals and plants portrayed in an abstract but still discernible way. That evening, in the small village hall, I felt two pieces of my life—dance and field biology—merge, at last, into one.

I also recognized the power of artists to communicate how they feel about their subjects to an audience. Although ecologists and environmental scientists possess a tremendous amount of information about particular animals, plants, and interactions, our training dictates that we leave our emotions out of the telling. Through my exchanges with these dancers and other artists, I learned ways of augmenting conservation efforts. In that small auditorium in Monteverde, people seemed far more inspired than I had ever seen them at a scientist's lecture or conservationist workshop. Nevertheless, I knew that my "science-y" input into the dance had added something. My time in the forest with the dancers, pointing out the details of interactions—of bird and flower, root and soil, vine and supporting tree—that I understood from systematic, long-term observation gave greater accuracy and depth to their movements. In the end, all of us were fascinated by how we managed to overlap and merge our fields. Science and art have much in common, including specialized tools and techniques, the patience and discipline to undertake repetitive tasks, and the need for a critical audience. At the same time, we share important desires: to enter that quiet state of mind in which our attention is wholly focused on what is right around us, and to awaken a feeling of connection and appreciation in our audiences. Perhaps partnerships between scientists and artists such as the one I enjoyed with Capacitor will create a new way of doing conservation in the future.

I wondered if this interdisciplinary approach to merge perspectives would incur risks as well as benefits, a pas de deux of pluses and minuses. In choosing to make trees and forests more understandable to nonscientists, would I somehow risk losing the depth that comes from working exclusively within my discipline, or losing the time I need to generate scientific papers for my academic colleagues? So far, I haven't felt my brain or my academic reputation slipping. Even if I had, I believe the risk of holding science as a mysterious realm that laypeople cannot enter is far more serious, especially during these times when it is becoming ever more critical that we—all of us—understand as much as we can about natural processes.

In the Monteverde forest—and in this book—I have learned much by combining science, art, and the humanities. I find that mixing these approaches has helped me to better grasp how intimately trees and humans are intertwined. Trees fulfill needs at all levels, from the basics of food and shelter and health, to our sense of time and play, to our expressions of creativity and spirituality.

Trees are complex, beautiful, fragile, and strong. Because of these characteristics, I think of them as ambassadors for the rest of nature. Trees also have the cachet of their age and size, as well as their usefulness to humans and cultural meaningfulness. For those reasons, trees need our understanding and protection. Every individual has a role to play, be it as a homeowner planting a tree in his backyard, a prisoner tending a mesh bag of moss, or a politician signing carbon-trading legislation into law.

MINDFUL RELATIONSHIPS

Bill Yake, a friend of mine and a poet, sent me a poem in which he tried to describe in simple verbs what trees do.

The Tree as Verb

"The true formula for thought is: The cherry tree is all that it does."
—*Ernest Fenollosa*

seed, swell, press and push, sprout, bud, curl, bloom, unfurl, quicken,
ripen, and dispense.
remain.
blotch, ferment, rot and mushroom.
germinate.
probe, grope, root, draw in, draw up, dole out, absorb, allot, assimilate,
respire, reconstitute, release.
senesce.
reach, brace, resist, avoid, deflect, split, notch, rustle, shake, bend and
shimmy.
occupy.
cover, mask, obscure, protect, enclose and hide; tolerate, support, feed,
shade, harbor and disguise.
stand, sketch out, stretch out, fork, reach, branch, divide, incline and sway.
reclaim, endure and burn.
return, leaf out, green, synthesize, digest, night-quiver, yellow, wilt and
wither, abscise, and again give way.

Bill noted that more verbs might need to be added as awareness of trees expands, and he invited me to add any that he had missed. I sent him three more:

Symbolize
Shelter
Be

I end this book not with science, nor with politics, but rather with a poem about a relationship. For relationships are what move and motivate humans toward action, toward change, and toward mindfulness. In this poem, as in so many of the relationships between trees and humans, the line among all of us becomes so blurred that I wonder: of the woman, man, tree, and earth, where does one end and the other begin?

The Tree

And when we woke it was like nothing
Ever dreamt before this: wrist, neck,
The hollow behind the knee, your hair

Filling my hands, all of it while we turned
And turned until we were unforgivable,
Adamant with bark, as if a wayward god had come

Upon us, bewitching breast to breast, fingers
Still tracing a vein, a thigh
No longer intent on destination

But in the keep of one limb resting on another, breath
Lingering in the leaves, at the edge of a road
Where we were once lost, your hand faithful

In its nest, your mouth on my mouth
Caught, our feet tangled, looking for earth.

—Sophie Cabot Black

SCIENTIFIC NAMES OF TREES IN THE TEXT

acacia	*Robinia* spp.
achiote	*Bixa orellana*
allspice	*Pimenta officinalis*
almond	*Prunus dulcis*
alpine spruce	*Picea abies*
American ash	*Fraxinus americana*
American elm	*Ulmus americana*
apple	*Malus* spp.
apricot	*Prunus armeniaca*
ash	*Fraxinus* spp.
atlas cedar	*Cedrus* spp.
avocado	*Persea americana*
balata	*Manilkara bidentata*
bald cypress	*Taxodium distichum*
balsa	*Ochroma* spp.
banyan	*Ficus benghalensis*
baobab	*Adansonia digitata*
bay laurel	*Laurus nobilis*
beech	*Fagus sylvatica*
Bermuda cedar	*Juniperus bermudiana*
bigleaf maple	*Acer macrophyllum*
birch	*Betula* spp.
black maple	*Acer nigrum*

black oak	*Quercus velutina*
black pepper	*Piper nigrum*
black spruce	*Picea mariana*
black walnut	*Juglans nigra*
blackjack oak	*Quercus dumosa*
blue spruce	*Picea pungens*
Bodhi	*Ficus religiosa*
bristlecone pine	*Pinus longaeva*
cacao	*Theobroma cacao*
California redwood	*Sequoia sempervirens*
camellia	*Camellia* spp.
canoe birch	*Betula papyrifera*
carob	*Ceratonia siliqua*
casuarina	*Casuarina* spp.
cedar of Lebanon	*Cedrus libani*
ceiba	*Ceiba* spp.
cherry	*Prunus* spp.
cinchona	*Cinchona officinalis*
cinnamon	*Cinnamomum verum zeylanicum*
citron	*Citrus medica*
clove	*Syzygium aromaticum*
coast redwood	*Sequoia sempervirens*
coconut palm	*Cocos nucifera*
common fig	*Ficus carica*
common hazel	*Corylus avellana*
cork oak	*Quercus suber*
cottonwood	*Populus trichocarpa*
cypress	*Chamaecyparis* spp.
date palm	*Phoenix dactylifera*
dawn redwood	*Metasequoia glyptostroboides*
dogwood	*Cornus* spp.
Douglas-fir	*Pseudotsuga menziesii*
eastern red cedar	*Juniperus virginiana*
eastern white pine	*Pinus strobus*
ebony	*Diospyros ebenum*
elephant	*Bursera microphylla*
elm	*Ulmus* spp.
Engelmann spruce	*Picea engelmannii*

English oak	*Quercus robur*
English yew	*Taxus baccata*
eucalyptus	*Eucalyptus* spp.
euphorbia	*Euphorbia* spp.
European beech	*Fagus sylvatica*
fig	*Ficus* spp.
fir	*Abies* spp.
frankincense	*Boswellia sacra*
giant arborvitae	*Thuja plicata*
giant sequoia	*Sequoia giganteum*
ginkgo	*Ginkgo biloba*
grapefruit	*Citrus × paradisi*
guanacaste	*Enterolobium cyclocarpum*
gutta percha	*Palaquium gutta*
Hawaiian koa	*Acacia koa*
hawthorn	*Crataegus* spp.
hazelnut	*Corylus* spp.
hickory	*Carya* spp.
holly	*Ilex aquifolium*
horse-chestnut	*Aesculus hippocastanum*
incense-cedar	*Calocedrus decurrens*
inga	*Inga* spp.
Italian cypress	*Cupressus sempervirens*
jackfruit	*Artocarpus heterophyllus*
Japanese cedar	*Cryptomeria japonica*
juniper	*Juniperus* spp.
koa	*Acacia koa*
kola	*Kola vera*
larch	*Larix laricina*
laurel	*Laurus nobilis*
lemon	*Citrus × lemon*
lime	*Citrus* spp.
linden	*Tilia americana*
live oak	*Quercus virginiana*
loblolly pine	*Pinus taeda*
lodgepole pine	*Pinus contorta*
longleaf pine	*Pinus palustris*
madrone	*Arbutus menziesii*

mahogany	*Swietenia macrophylla*
mango	*Mangifera indica*
maple	*Acer* spp.
mastic	*Pistacia lentiscus*
mesquite	*Prosopis* spp.
mulberry	*Morus* spp.
myrrh	*Commiphora myrrha*
myrtle	*Myrtus communis*
neem	*Azadirachta indica*
Norfolk Island pine	*Araucaria heterophylla*
northern hard maple	*Acer saccharum*
nutmeg	*Myristica moschata*
oak	*Quercus* spp.
olive	*Olea europaea*
orange	*Citrus* × *sinensis*
Pacific yew	*Taxus brevifolia*
palm	plant family Arecaceae
palmetto	*Sabal* spp.
papaya	*Carica papaya*
peach	*Prunus persica*
pear	*Pyrus communis*
peelu	*Salvadora persica*
persimmon	*Diospyros virginiana*
pine	*Pinus* spp.
pistachio	*Pistacia vera*
plum	*Prunus* spp.
ponderosa pine	*Pinus ponderosa*
poplar	*Populus* spp.
post oak	*Quercus dumosa, Q. stellata*
privet	*Ligustrum* spp.
quaking aspen	*Populus tremuloides*
red alder	*Alnus rubra*
red maple	*Acer rubrum*
redwood	*Sequoia sempervirens*
river birch	*Betula nigra*
rosewood	*Aniba rosaeodora*
sal	*Shorea robusta*
sandalwood	*Santalum* spp.

sapodilla	*Manilkara zapota*
sausage	*Kigelia africana*
scrubby oak	*Quercus dumosa*
silk-cotton	*Ceiba* spp.
silver maple	*Acer saccharinum*
Sitka spruce	*Picea sitchensis*
slash pine	*Pinus elliottii*
southern magnolia	*Magnolia grandiflora*
Spanish cedar	*Cedrela odorata*
spruce	*Picea* spp.
star fruit	*Averrhoa carambola*
stilt palm	*Socratea exorrhiza*
strangler fig	*Ficus* spp.
sugar maple	*Acer saccharum*
sweet cherry	*Prunus avium*
sycamore	*Platanus* spp.
sycamore fig	*Ficus sycomorus*
tabonuco	*Dacryodes excelsa*
tamarind	*Tamarindus indica*
tamarisk	*Tamarix* spp.
tangerine	*Citrus reticulata*
teak	*Tectona grandis*
terebinth	*Pistacia terebinthus*
tropical fig	*Ficus tuerckheimii*
umbrella acacia	*Acacia tortilis*
valley oak	*Quercus lobata*
virola	*Virola* spp.
walnut	*Juglans* spp.
weeping willow	*Salix × sepulcralis*
western hemlock	*Tsuga heterophylla*
western red cedar	*Thuja plicata*
white ash	*Fraxinus americana*
white oak	*Quercus alba*
willow	*Salix* spp.
witch hazel	*Hamamelis* spp.
yellow pine	*Pinus elliottii*, *P. ponderosa*, *P. taeda*
yew	*Taxus* spp.

NOTES

INTRODUCTION

Epigraph: Michael S. Glaser, "A Blessing for the Woods," *Christian Science Monitor*, May 19, 1999.

p. 14 *"When we have learned how to listen to trees":* Hermann Hesse, "On Trees," in *Wandering: Notes and Sketches*, translated by James Wright (New York: Farrar, Straus, and Giroux, 1972; © Frankfurt am Main: Suhrkamp, 2005), 59.

CHAPTER 1. WHAT IS A TREE?

Epigraph: Hermann Hesse, "On Trees," in *Wandering: Notes and Sketches*, translated by James Wright (New York: Farrar, Straus, and Giroux, 1972; © Frankfurt am Main: Suhrkamp, 2005), 58.

p. 22 *This complex process of photosynthesis:* "Photosynthesis," in *World of Biology*, ed. Kimberley A. McGrath (Detroit: Gale Group, 1999), 600.

p. 22 *Their needles can carry out photosynthesis:* D. Y. Hollinger, "Canopy Organization and Foliage Photosynthetic Capacity in a Broad-Leaved Evergreen Montane Forest," *Functional Ecology* 3 (1989): 53–62.

p. 24 *Some of their nutrient needs:* N. M. Nadkarni and T. Matelson, "Litter Dynamics within the Canopy of a Neotropical Cloud Forest, Monteverde, Costa Rica," *Ecology* 72 (1991): 2071–82.

p. 25 *Recent experimental research by Martin Freiberg:* Martin Freiberg, "The Influence of Epiphyte Cover on Branch Temperature in a Tropical Tree," *Plant Ecology* 153 (2001): 241–250.

p. 26 *A team of German and Swiss researchers recently documented:* Sabine Stuntz, Ulrich Simon, and Gerhard Zotz, "Rainforest Air-Conditioning: The

Moderating Influence of Epiphytes on the Microclimate in Tropical Tree Crowns," *International Journal of Biometeorology* 46 (2002): 53–59.

p. 26 *In the late 1970s, research into the ecological roles of epiphytes:* W. C. Denison, "Life in Tall Trees," *Scientific American* 228 (1973): 74–80; G. C. Carroll, L. H. Pike, J. R. Perkins, and M. A. Sherwood, "Biomass and Distribution Patterns of Conifer Twig Microepiphytes in a Douglas Fir Forest," *Canadian Journal of Botany* 58 (1980): 624–630.

p. 27 *In Monteverde, Costa Rica, my students and I have measured the amounts:* K. L. Clark, N. M. Nadkarni, D. Schaefer, and H. L. Gholz, "Atmospheric Deposition and Net Retention of Ions by the Canopy in a Tropical Montane Forest, Monteverde, Costa Rica," *Journal of Tropical Ecology* 14 (1998): 27–45; K. L. Clark, N. M. Nadkarni, and H. L. Gholz, "Growth, Net Production, Litter Decomposition, and Net Nitrogen Accumulation by Epiphytic Bryophytes in a Tropical Montane Forest," *Biotropica* 30 (1998): 12–23.

p. 27 *this soil can amount to as much as 280 pounds:* N. M. Nadkarni, "Nutrient Capital of Canopy Epiphytes in an *Acer macrophyllum* Community, Olympic Peninsula, Washington State," *Canadian Journal of Botany* 77 (1985): 136–142.

p. 28 *Among the many new insights he gleaned:* M. R. Keyes and C. C. Grier, "Above- and Below-Ground Net Production in 40-Year-Old Douglas-Fir Stands on Low and High Productivity Sites," *Canadian Journal of Forest Research* 11 (1981): 599–605.

p. 29 *In 2001, scientists at the Environmental Protection Agency:* M. G. Johnson, D. T. Tingey, D. L. Phillips, and M. J. Storm, "Advancing Fine Root Research with Minirhizotrons," *Environmental and Experimental Botany* 45 (2001): 263–289.

p. 30 *"A tree communicates":* Howard McCord, *Walking Edges: A Book of Obsessional Texts* (Rossford, Ohio: Raincrow Press, 1982), 38–39.

p. 30 *"The tree is more than first a seed":* Antoine de Saint-Exupéry, *The Wisdom of the Sands*, translated from the French by Stuart Gilbert (New York: Harcourt Brace and Co., 1950), 22.

p. 31 *In the 1980s, Claus Mattheck:* Claus Mattheck, *Design in Nature: Learning from Trees*, trans. William Linnard (Berlin: Springer-Verlag, 1998).

p. 34 *In 1908, U.S. Forest Service botanist G. B. Sudworth:* G. B. Sudworth, *Forest Trees of the Pacific Slope* (Washington, D.C.: Government Printing Office, 1908).

p. 35 *A one-acre tropical rainforest plot in Amazonian Ecuador:* H. Balslev et al., "Species Count of Vascular Plants in One Hectare of Humid Lowland Forest in Amazonian Ecuador," in *Forest Biodiversity in North, Central, and South America, and the Caribbean,* ed. F. Dallmeier and J. A. Comisky. (Paris: UNESCO, 1998), 2:585–594; R. Primack and R. Corlett, *Tropical Rainforests: An Ecological and Biogeographical Comparison* (Malden, Mass.: Blackwell Science, 2005).

p. 36 *an adjunct curator at Chicago's Field Museum:* http://fm2.fieldmuseum.org/plantguides; Kyle E. Harms, Richard Condit, Stephen P. Hubbell, and Robin B. Foster, "Habitat Associations of Trees and Shrubs in a 50-ha Neotropical Forest Plot," *Journal of Ecology* 89 (2001): 947–959; S. P. Hubbell, J. A. Ahumada, R. Condit, and R. B. Foster, "Local Neighborhood Effects on Long-Term Survival of Individual Trees in a Neotropical Forest," *Ecological Research* 16 (2001): 859–875.

p. 38 *a booklet depicting the shapes that occur in orchards:* Devon County Council Environment Department, "Orchards—A Practical Guide for Management," 1999 (loose-leaf pack of 12 pages); Countryside Council for Wales, "Traditional Orchards," 1995.

p. 39 *devised a system called "tree architecture":* F. Hallé, R. A. A. Oldeman, and P. B. Tomlinson, *Tropical Trees and Forests: An Architectural Analysis* (Berlin: Springer-Verlag, 1978).

p. 40 *a new and integrative conceptual framework to categorize trees:* Judith B. Cushing, Anne C. S. Fiala, Nalini M. Nadkarni, and Lee C. Zeman, "Semantics for Integrating Forest Structure Data," talk presented at the 15th Annual International Conference on Intelligent Systems for Molecular Biology (ISMB) and 6th European Conference on Computational Biology (ECCB), Vienna, Austria, July 21–25, 2007.

p. 42 *Today, about 30 percent of the world's land surface:* Food and Agriculture Organization of the United Nations, *Global Forest Resources Assessment 2005,* FAO Forestry Paper 147, Rome. Available at ftp://ftp.fao.org/docrep/fao/008/A0400E/A0400E00.pdf and ftp://ftp.fao.org/docrep/fao/008/A0400E/A0400E14.pdf (tables).

p. 42 *Using such diverse sources as tree counts . . . and remote imagery:* Global Forest Resources Assessment 2005. FAO Forestry Paper 147, Rome, available as ftp://ftp.fao.org/docrep/fao/008/A0400E/A0400E00.pdf and ftp://ftp.fao.org/docrep/fao/008/A0400E/A0400E14.pdf (tables).

p. 43 *Geoffrey (Jess) Parker, an ecologist:* G. G. Parker, "Structure and Microclimate of Forest Canopies," in *Forest Canopies*, ed. Margaret D. Lowman and Nalini M. Nadkarni (San Diego, Calif.: Academic Press, 1995), 73–106.

p. 44 *quantitatively describes the airspace of world forests:* Roman J. Dial, Stephen C. Sillett, Marie E. Antoine, and Jim C. Spickler, "Methods for Horizontal Movement through Forest Canopies," *Selbyana* 25 (2004): 151–163.

p. 46 *world's tallest living thing:* Robert Van Pelt, *Forest Giants of North America* (Seattle: University of Washington Press, 2001); Michael G. Barbour, Marjorie Popper, and John Evarts, eds., *Coast Redwood: A Natural and Cultural History* (Los Olivos, Calif.: Cachuma Press, 2001); *Sequoia sempervirens* in the "Gymnosperm Database," www.conifers.org/cu/se/index.htm.

p. 46 *trees that "stand alone":* Hesse, "On Trees," 57.

p. 47 *on a scientific project measuring throughfall:* N. M. Nadkarni and M. Sumera, "Old-Growth Forest Canopy Structure and Its Relationship to Throughfall Interception," *Forest Science* 50 (2004): 290–298.

p. 48 *When epiphytes fall to the ground:* T. J. Matelson, N. M. Nadkarni, and J. T. Longino, "Survivorship of Fallen Epiphytes in a Neotropical Cloud Forest, Monteverde, Costa Rica," *Ecology* 74 (1993): 265–269.

p. 50 *Documenting the dynamics of the stand:* T. J. Matelson, N. M. Nadkarni, and R. Solano, "Tree Damage and Annual Mortality in a Montane Forest in Monteverde, Costa Rica," *Biotropica* 27 (1995): 441–447.

p. 51 *thought-provoking three-dimensional installations:* www.brucechao.net.

p. 52 *Robert Horton and Arthur Strahler established a classification system:* Robert E. Horton, "Erosional Development of Streams and Their Drainage Basins," *Geological Society of America Bulletin* 45 (1945): 275–370; Arthur Strahler, "Quantitative Analysis of Watershed Geomorphology," *Transactions of the American Geophysical Union* 38 (1957): 913–920.

p. 54 *Using Horton and Strahler's methods, researchers determined:* D. J. Barnes and C. J. Crossland, "Diurnal and Seasonal Variations in the Growth of a Staghorn Coral Measured by Time-Lapse Photography," *Limnological Oceanography* 25 (1980): 1113–1117.

p. 55 *Recently, my family and I visited the "Bodies" exhibit:* The website for the traveling exhibit is www.bodiestheexhibition.com.

p. 56 *visual-imaging scientists Gert van Tonder and Michael Lyons:* Gert J. van Tonder and Michael J. Lyons, "Visual Perception in Japanese Rock Garden Design," *Axiomathes* 15 (2005): 353–371.

p. 56 *The Tree of Life:* www.tolweb.org/tree.

CHAPTER 2. GOODS AND SERVICES

Epigraph: Marcel Proust, *Pleasures and Regrets*, translated by Louise Varese (London: Peter Owen, 2001), 165.

p. 60 *[a cord] of wood produces:* North Carolina Forestry Association, www.ncforestry.org/docs/Products/cord.htm.

p. 61 *A cut log arriving at the mill:* William Dietrich, *The Final Forest: The Battle for the Last Great Trees of the Pacific Northwest* (New York: Penguin, 1993); Jim L. Bowyer, Rubin Shmulsky, and John G. Haygreen, *Forest Products and Wood Science: An Introduction*, fifth edition (Ames, IA: Iowa State University Press, 2007).

p. 63 *plywood:* National Council of the Paper Industry for Air and Stream Improvement, Technical Bulletin No. 503, www.ecy.wa.gov/programs/air/pdfs/plywood3.pdf.

p. 65 *Other secondary compounds:* www.nslc.wustl.edu/courses/Bio343A/2005/secondary.pdf.

p. 65 *A subset of terpenes, called limonoids:* S. Hasegawa and M. Miyake, "Biochemistry and Biological Functions of Citrus Limonoids," *Food Review International* 12 (1996): 413; D. Champagne, E.K. Opender, and B.I. Murray, "Biological Activity of Limonoids from the Rutales," *Phytochemistry* 31 (1992): 377–394.

p. 66 *In 2001, an estimated two billion pallets:* J. Clarke, "Pallets 101: Industry Overview and Wood, Plastic, Paper, and Metal Options," www.ista.org/Knowledge/Pallets_101-Clarke_2004.pdf.

p. 66 *These in turn create other problems:* Kim Todd, *Tinkering with Eden: A Natural History of Exotics in America* (New York: W.W. Norton, 2001); U.S.D.A. Foreign Agriculture Service, "AgExporter," http://www.fas.usda.gov/info/agexporter/2001/jan/PackingRegulations.htm.

p. 66 *"Paper or plastic?":* Institute for Lifecycle Environmental Assessment, "Paper vs. Plastic Bags, www.ilea.org/lcas/franklin1990.html; "Resource and Environmental Profile Analysis of Polyethylene and Unbleached Paper Grocery Sacks," www.americanchemistry.com/s_plastics/bin.asp?cid=1211&did=4601&doc=file.pdf.

p. 67 *Some secondary compounds are attractive to pollinators:* Michael Pollan, *The Botany of Desire: A Plant's Eye View of the World* (New York: Random House, 2001).

p. 67 *The humble pack of chewing gum:* Michael Redclift, *Chewing Gum: The Fortunes of Taste* (New York: Routledge, 2004).

p. 68 *artificial vanillin was produced:* L. J. Esposito et al., "Vanillin," in *Kirk-Othmer Encyclopedia of Chemical Technology,* 4th ed. (New York: John Wiley, 1997), 812–825; M. B. Hocking, "Vanillin: Synthetic Flavoring from Spent Sulfite Liquor," *Journal of Chemical Education* 74 (1997): 1055; G. M. Lampman et al., "Preparation of Vanillin from Eugenol and Sawdust," *Journal of Chemical Education* 54 (1977): 776–778.

p. 69 *Other tree-derived goods include spices:* Tony Hill, *The Contemporary Encyclopedia of Herbs and Spices: Seasonings for the Global Kitchen* (New York: John Wiley, 2004).

p. 71 *"There is in some parts of New England":* Robert Boyle, *Philosophical Works* (1663), quoted in *Encyclopedia Americana* (1987), 416–417.

p. 71 *syrup-making processes:* "How to Tap Maple Trees and Make Maple Syrup," University of Main Extension Bulletin #7036, www.umext.maine.edu/onlinepubs/PDFpubs/7036.pdf.

p. 72 *In 2005, over 47,000 tons of syrup:* "New England Agricultural Statistics," U.S. Department of Agriculture, National Agricultural Statistics Service, www.nass.usda.gov/nh/mapleconf2005.pdf.

p. 72 *The cause of maple sap flow:* Theodore T. Kozlowski and Stephen G. Pallardy, *Physiology of Woody Plants,* 2d ed. (New York: Academic Press, 1997); J. W. Martin, "The Physiology of Maple Sap Flow," in *The Physiology of Forest Trees,* ed. Kenneth V. Thimann (New York: Ronald Press, 1958).

p. 73 The Story of Ferdinand: Munro Leaf and Robert Lawson, *The Story of Ferdinand* (New York: Viking, 1936).

p. 74 *The cork oak:* F. H. Schweingruber, A. Börner, and E.-D. Schulze, *Atlas of Woody Plant Stems: Evolution, Structure, and Environmental Modifications* (Berlin: Springer-Verlag, 2006); Katherine Esau, *Plant Anatomy,* 2d ed. (New York: John Wiley, 1965); William C. Dickison, *Integrative Plant Anatomy* (Orlando, Fla.: Academic Press, 2000). General references for cork harvesting and the cork industry: George M. Taber, *To Cork or Not to Cork: Tradition, Romance, Science, and the Battle for the Wine Bottle* (New York: Scribner, 2007); M. C. Varela, "Cork and the Cork Oak System" (1999), www.fao.org/docrep/x1880e/x1880e08.htm#cork%20and%20the%20cork%20oak%20system; Virginia Tech Forestry Department, "Cork Oak," www.cnr.vt.edu/DENDRO/dendrology/syllabus2/factsheet.cfm?ID=553.

p. 78 *disposable wooden chopsticks:* "Shortage of Chopsticks Worries Japan as China

Goes Ecological," www.chinadaily.com.cn/china/2006-05/15/content_589763.htm; Mari Yamaguchi, "Deforestation Fears Spur China to Tax Chopsticks," *Chicago Sun-Times*, May 14, 2006, accessed at http://findarticles.com/p/articles/mi_qn4155/is_20060514/ai_n16366636.

p. 78 *Tree gum:* Marc Paye, Andre O. Barel, and Howard I. Maibach, *Handbook of Cosmetic Science and Technology*, 2d ed. (Boca Raton, Fla.: Informa Healthcare, 2005); Ruth Winter, *A Consumer's Dictionary of Cosmetic Ingredients: Complete Information about the Harmful and Desirable Ingredients Found in Cosmetics and Cosmeceuticals* (New York: Three Rivers Press, 2005).

p. 80 *Turpentine is tapped from pine trees:* Luther Burbank, *Trees Whose Products Are Useful Substances: From the Sugar Maple to the Turpentine Tree* (Athena University Press, 2004); Food and Agriculture Organization of the United Nations, *Gum Naval Stores: Turpentine and Rosin from Pine Resin*, available at http://www.fao.org/docrep/V6460E/v6460e04.htm.

p. 81 *the camphor laurel:* www.3dchem.com/molecules.asp?ID=203.

p. 81 *"Many people . . . contribute to the making of a book":* Roy Rada and Richard Forsyth, *Machine Learning: Applications in Expert Systems and Information Retrieval* (New York: John Wiley and Sons, 1986), 12.

p. 82 *How much paper* do *we use?:* Food and Agriculture Organization, *State of the World's Forests 2007* (Rome, 2007), available at www.fao.org/docrep/009/a0773e/a0773e00.htm.

p. 83 *The first mass-produced pencils:* Henry Petroski, *The Pencil: A History of Design and Circumstance* (New York: Alfred A. Knopf, 1989).

p. 84 *a traditional casket:* Casket and Funeral Supply Association, www.cfsaa.org/about.php.

p. 84 *Memorial Ecosystems:* www.memorialecosystems.com.

p. 85 *"Brilliantly red and orange flames":* http://realtravel.com/kathmandu-journals-j2022960.html.

p. 87 *In 2007, the Parks Department of New York City:* David K. Randall, "Maybe Only God Can Make a Tree, but Only People Can Put a Price on It," *New York Times*, April 18, 2007, www.nytimes.com/2007/04/18/nyregion/18trees.html.

p. 87 *The program estimates such factors:* Y. Tetsuya, H. Kikuo, N. Yoshiteru, and M. Hirohiko, "Characteristics of Transpiration of Revegetated Trees of Urban Space in Summer Season," *Forest Resources and Environment* 39 (2001): 1–18.

CHAPTER 3. SHELTER AND PROTECTION

Epigraph: John Clare, "In Hilly-wood," *The Rural Muse* (London: Whittaker, 1835).

p. 89 *"Tree houses inspire dreams":* Peter Nelson, *Treehouses: The Art and Craft of Living Out on a Limb* (Boston: Houghton Mifflin, 1994), 129.

p. 90 *Out 'n' About Treesort:* www.treehouses.com/treehouse/treesort/home.html.

p. 93 *Yayoi dwellings of ancient Japan:* J. Maringer, "Dwellings in Ancient Japan: Shapes and Cultural Context," *Asian Folklore Studies* 39 (1980): 115–123.

p. 93 *"Sweat lodges were usually domed":* Mikkel Aaland, "Native American Sweat Lodge: History of Sweat Lodges," available at http://www.cyberbohemia.com/Pages/historysweatlod.htm.

p. 94 *To build a typical log cabin:* "Log Cabin Log Count," www.ankn.uaf.edu/publications/VillageMath/logcabin.html; David B. McKeever and Robert B. Phelps, "Wood Products Used in New Single-Family House Construction: 1950 to 1992," *Forest Products Journal* 44 (1994): 66–74, available at www.fpl.fs.fed.us/documnts/pdf1994/mckee94a.pdf.

p. 95 *High latitude and high-altitude trees must adapt:* Stephen F. Arno and Ramona P. Hammerly, *Timberline: Mountain and Arctic Forest Frontiers* (Seattle: Mountaineers Books, 1984); J. L. Hadley and W. K. Smith, "Influence of Krummholz Mat Microclimate on Needle Physiology and Survival," *Oecologia* 73 (1987): 82–90; G. Stevens and J. Fox, "The Causes of Treeline," *Annual Review of Ecology and Systematics* 32 (1991): 177–191.

p. 96 *How do desert trees cope?:* Gerald E. Wickens, *Ecophysiology of Economic Plants in Arid and Semi-Arid Lands* (New York: Springer, 1998); Gerald E. Wickens and Pat Lowe, *The Baobabs: Pachycauls of Africa, Madagascar and Australia* (New York: Springer, 2008); Thomas Pakenham, *The Remarkable Baobab* (New York: Norton, 2004).

p. 102 *About 25 percent of the world's fuel wood:* Roger Sands, *Forestry in a Global Context* (Oxford: CABI Publishers, 2005).

p. 102 *Monteverde Conservation League:* www.mclus.org.

p. 104 *windbreaks or shelterbelts:* P. R. Bird, "Tree Windbreaks and Shelter Benefits to Pasture in Temperate Grazing Systems," *Agroforestry Ecosystems* 41 (2004): 35–54; R. W. Gloyne, "Some Effects of Shelterbelts upon Local and Micro Climate," *Forestry* 27 (1954): 85–95.

p. 105 *shade-grown coffee:* www.equalexchange.com; www.transfairusa.org.

p. 107 *the importance of city parks:* Chris Walker, "The Public Value of Urban Parks," Urban Institute/Wallace Foundation, 2004, www.wallacefoundation.org/NR/rdonlyres/5EB4590E-5E12-4E72-B00D-613A42E292E9/0/ThePublicValueofUrbanParks.pdf.

p. 107 *a venue where youth development can be fostered:* Committee on Community-Level Programs for Youth, Jacquelynne Eccles and Jennifer Appleton Gootman, eds., National Research Council and Institute of Medicine, *Community Programs to Promote Youth Development* (Washington D.C.: National Academies Press, 2002).

p. 107 *Empowering Youth Initiative:* Margery Austin Turner, "Urban Parks as Partners in Youth Development," Urban Institute/Wallace Foundation, 2004, www.wallacefoundation.org/NR/rdonlyres/55C091D5-5EB0-4798-81B2-621975E016A8/0/UrbanParksasPartnersinYouthDevelopment.pdf.

p. 108 *parklike public spaces:* F. Kuo, M. Bacaicoa, and W. C. Sullivan, "Transforming Inner-City Landscapes: Trees, Sense of Safety, and Preference," *Environment and Behavior* 30 (1998): 28–59.

p. 109 *"I rock high in the oak":* William Stafford, "Little Rooms," in *The Way It Is: New and Selected Poems* (Saint Paul, Minn.: Graywolf Press, 1998), 210.

CHAPTER 4. HEALTH AND HEALING

Epigraph: Hermann Hesse, "On Trees," in *Wandering: Notes and Sketches*, translated by James Wright (New York: Farrar, Straus, and Giroux, 1972; © Frankfurt am Main: Suhrkamp, 2005), 58.

p. 113 *the treatment of ailments ranging from diabetes to depression:* G. Prasad and M. V. Reshmi, *Manual of Medicinal Trees* (Jodhpur, India: Agrobios, 2003); Bryan Abbot Hanson, *Understanding Medicinal Plants: Their Chemistry and Therapeutic Action* (Binghamton, N.Y.: Haworth Herbal Press, 2005); Ben-Erik van Wyk and Michael Wink, *Medicinal Plants of the World* (Portland, Oreg.: Timber Press, 2004).

p. 115 *One of the participants, Mark Plotkin:* Mark J. Plotkin, *Tales of a Shaman's Apprentice: An Ethnobotanist Searches for New Medicines in the Amazon Rain Forest* (New York: Penguin Books, 1994); Mark J. Plotkin, *Medicine Quest: In Search of Nature's Healing Secrets* (New York: Viking Penguin, 2001).

p. 116 *the practice of "bioprospecting":* Rashid Hassan, Robert Scholes, and Neville Ash, eds., *Ecosystems and Human Well-Being: Current State and Trends—*

Findings of the Condition and Trends Working Group of the Millennium Ecosystem Assessment (Washington, D.C.: Island Press, 2005); Michael J. Balick, Elaine Elisabetsky, and Sarah A. Laird, eds., *Medicinal Resources of the Tropical Forest and Its Importance to Human Health* (New York: Columbia University Press, 1996).

p. 117 *Will Setzer, a chemist, was collaborating with an ecologist, Bob Lawton:* R. O. Lawton and W. N. Setzer, "Floral Essential Oil of *Guettarda poasana* Inhibits Yeast Growth," *Biotropica* 25 (1993): 483–486; D. M. Moriarty et al., "Lupeol Is the Cytotoxic Principle in the Leaf Extract of *Dendropanax* cf. *querceti*," *Planta Medicine* 64 (1998): 370–372.

p. 118 *Tropical medicinal plants play particularly important roles:* Michael J. Balick and Paul Alan Cox, *Plants, People, and Culture: The Science of Ethnobotany* (New York: Scientific American Library, 1996).

p. 118 *Researchers from the New York Botanical Garden:* Michael J. Balick and Robert Mendelsohn, "Assessing the Economic Value of Traditional Medicines from Tropical Rain Forests," *Conservation Biology* 6 (1992): 128–130.

p. 119 *In the Great Lakes region, the Ojibwa:* Daniel E. Moerman, *Native American Ethnobotany* (Portland, Oreg.: Timber Press, 1998).

p. 121 *In a landmark study published in the journal* Science*:* R. S. Ulrich, "View Through a Window May Influence Recovery from Surgery," *Science* 224 (1984): 420–421; R. S. Ulrich, "Visual Landscapes and Psychological Well-Being," *Landscape Research* 4 (1979): 17–23.

p. 121 *Other studies showed that environments with nature-related imagery:* P. Leather, M. Pyrgas, D. Beale, and C. Lawrence, "Windows in the Workplace: Sunlight, View, and Occupational Stress," *Environment and Behavior* 30 (1998): 739–762. S. Verderber, "Dimensions of Person-Window Transactions in the Hospital Environment," *Environment and Behavior* 18 (1986): 450–466.

p. 122 *environments enhanced by plants and nature imagery:* J. H. Heerwagen and G. H. Orians, "Adaptations to Windowlessness: A Study of the Use of Visual Decor in Windowed and Windowless Offices," *Environment and Behavior* 18 (1986): 623–639.

p. 122 *SkyFactory:* www.theskyfactory.com.

p. 122 *Sociological research by Virginia Lohr and Caroline Pearson-Mims:* Virginia I. Lohr and Caroline H. Pearson-Mims, "Responses to Scenes with Spreading, Rounded, and Conical Tree Forms," *Environment and Behavior* 38 (2006): 667–688.

p. 123 *the "savanna hypothesis":* Gordon H. Orians, "Habitat Selection: General Theory and Applications to Human Behavior," in *The Evolution of Human Social Behavior*, ed. Joan S. Lockard (Amsterdam: Elsevier, 1980), 49–66; Orians, "An Ecological and Evolutionary Approach to Landscape Aesthetics," in *Landscape Meanings and Values*, ed. Edmund C. Penning-Rowsell and David Lowenthal (London: Allen and Unwin, 1986), 3–22.

p. 124 *"I will remember that there is art to medicine":* "A Modern Hippocratic Oath," by Dr. Louis Lasagna, cited by Association of American Physicians and Surgeons, www.aapsonline.org/ethics/oaths.htm.

p. 125 *In a five-year study published in the* New England Journal of Medicine: H. J. Burstein, S. Gelber, E. Guadagnoli, and J. Weeks, "Use of Alternative Medicine by Women with Early-Stage Breast Cancer," *New England Journal of Medicine* 340 (1999): 1733–1739.

p. 127 *the literature on visualization therapy:* Mind-Body Medicine for Cancer, www.webmd.com/balance/features/mind-body-medicine-for-cancer; Dora Kunz and Dolores Krieger, *The Spiritual Dimension of Therapeutic Touch* (Rochester, Vt.: Bear & Co., 2004).

p. 127 *Empowered Within Project:* www.empoweredwithin.com/products/child_cancer.html.

p. 127 *"For our health, we have to preserve our physical world":* Moyra Caldecott, *Myths of the Sacred Tree* (Rochester, Vt.: Destiny Books, 1993), 2.

p. 128 *Healing Trees Project:* www.livingmemorialsproject.net.

p. 129 *A Window Between Worlds:* www.awbw.org.

p. 130 *"Throughout time, humans have looked to trees":* Gail Faith Edwards, *Opening Our Wild Hearts to the Healing Herbs* (Adirondack, N.Y.: Ash Tree, 2000), 32–33.

p. 133 *"the very process of restoring the land":* Chris Maser, *Forest Primeval: The Natural History of an Ancient Forest* (Corvallis: Oregon State University Press, 2001), 32.

p. 133 *"a vast intellectual legacy":* Edward O. Wilson, *Biophilia* (Cambridge, Mass.: Harvard University Press, 1984), 33.

CHAPTER 5. PLAY AND IMAGINATION

Epigraph: Stephen Nachmanovitch, *Free Play: Improvisation in Life and Art* (Los Angeles: J. P. Tarcher, 1990), 42.

p. 135 *We found that over one-third of the visits:* Nalini Nadkarni and Teri Matelson, "Bird Use of Epiphyte Resources in Neotropical Trees," *Condor* 91 (1989): 891–907.

p. 140 *In 1983, Peter "Treeman" Jenkins:* www.treeclimbing.com.

p. 142 *The hickory sticks wielded by Babe Ruth:* Teresa Riordan, "Just in Time for the Season: An Ash-Hickory-Maple Bat, a Plastic Glove and an Ultrasonic Ump," *New York Times*, May 15, 1995; Brendan I. Koerner, "As Laws Change, So Does the Baseball Bat," *New York Times*, July 1, 2007.

p. 142 *Metal bats, introduced in the 1970s:* R. M. Greenwald, L. H. Penna, and J. J. Crisco, "Differences in Batted Ball Speed with Wood and Aluminum Baseball Bats: A Batting Cage Study," *Journal of Applied Biomechanics* 17 (2001): 241–252; Mike Jenkins, *Materials in Sports Equipment* (Cambridge: Woodhead Publishing, 2003).

p. 144 *In Scotland, the earliest woods:* Charles McGrath and David McCormick, *The Ultimate Golf Book: A History and a Celebration of the World's Greatest Game* (New York: Houghton Mifflin, 2002).

p. 148 *women athletes who sustained knee ligament injuries:* O. E. Olsen et al., "Relationship between Floor Type and Risk of ACL Injury in Team Handball," *Scandinavian Journal of Medicine Science in Sports* 13 (2003): 299–314.

p. 148 *nearly 23 million square feet of maple sports flooring:* "Wood Sports Floors: Minimizing Damaging Effects," *Architectural Record*, http://archrecord.construction.com/resources/conteduc/archives/research/1_00_3.asp.

p. 149 *World Conker Championships:* www.worldconkerchampionships.com.

p. 149 *"steel can be cold":* www.ultimaterollercoaster.com.

p. 151 *International Tree Climbing Championship:* http://itcc.isa-arbor.com.

p. 152 *bonsai, a living art form:* Deborah R. Koreshoff, *Bonsai: Its Art, Science, History and Philosophy* (Portland, Oreg.: Timber Press, 1997).

p. 153 *Topiary, the art of sculpting shrubs and trees:* Barbara Gallup and Deborah Reich, *The Complete Book of Topiary* (New York: Workman, 1988).

p. 154 *Pliny the Younger:* Pierre de la Ruffinière du Prey, *The Villas of Pliny from Antiquity to Posterity* (Chicago: University of Chicago Press, 1995).

p. 154 *He created the "Tree Circus":* www.arborsmith.com/treecircus.html.

p. 156 *young people are losing contact with nature:* Robin C. Moore, "The Need for Nature: A Childhood Right," *Social Justice* 24 (1997): 203–220; John Beardsley, "Kiss Nature Goodbye: Marketing the Great Outdoors," *Harvard Design Magazine* 10 (2000), www.gsd.harvard.edu/research/publications/hdm/back/10beardsley.pdf; Marco Hüttenmoser, "Children and Their Living Surroundings: Empirical Investigations into the Significance of Living Surroundings for the Everyday Life and Development of Children," *Children's Environments* 12 (1995): 403–413.

p. 157 *In his book* Last Child*:* Richard Louv, *Last Child in the Woods: Saving Our*

Children from Nature-Deficit Disorder (Chapel Hill, N.C.: Algonquin Books, 2005).

p. 157 *Duke University Child Well-Being Index:* Child and Youth Well-Being Index, 2007 Report, available at www.childstats.gov/pubs.asp and www.soc .duke.edu/~cwi.

p. 157 *many studies show a relationship between lack of parks:* Ariane L. Bedimo-Rung, Andrew J. Mowen, and Deborah Cohen, "The Significance of Parks to Physical Activity and Public Health: A Conceptual Model," *American Journal of Preventive Medicine* 28 (2005): 159–168; Lawrence D. Frank and Peter O. Engelke, "The Built Environment and Human Activity Patterns: Exploring the Impacts of Urban Form on Public Health," *Journal of Planning Literature* 16 (2001): 202–218; Howard Frumkin, "Healthy Places: Exploring the Evidence," *American Journal of Public Health* 93 (2003): 1451–1456.

p. 158 *Louv . . . has been interviewed:* www.npr.org/templates/story/story.php ?storyId=4665933; www.grist.org/news/maindish/2006/03/30/louv/; www.post-gazette.com/pg/07137/786527-51.stm (*Pittsburgh Post-Gazette*).

p. 158 *Nature play could . . . emerge as a promising therapy:* John A. Hattie, Herbert W. Marsh, James T. Neill, and Garry E. Richards, "Adventure Education and Outward Bound: Out-of-Class Experiences That Make a Lasting Difference," *Review of Educational Research* 6 (1997): 43–87; Deborah Carlton Harrell, "Away from the Tube and into Nature, Children Find a New World," *Seattle Post-Intelligencer*, April 5, 2002, http://seattlepi .nwsource.com/local/65369_nature05.shtml.

p. 158 *IslandWood:* www.islandwood.org.

p. 158 *"Something will have gone out of us as a people":* Wallace Stegner's "Wilderness Letter," written in 1960 to David E. Pesonen of the Outdoor Recreation Resources Review Commission, www.wilderness.org/OurIssues/ Wilderness/wildernessletter.cfm.

p. 158 *In 2005, the Nature Conservancy of Washington:* P. McAllister, "An Inquiry in Support of Philanthropic Action for the Nature Conservancy of Washington, Seattle, Washington," unpublished report, 2003.

CHAPTER 6. CONNECTIONS TO TIME

Epigraph: Hermann Hesse, "On Trees," in *Wandering: Notes and Sketches*, translated by James Wright (New York: Farrar, Straus, and Giroux, 1972; © Frankfurt am Main: Suhrkamp, 2005), 57.

p. 167 *the . . . alpine spruce that Stradivari used for his instruments:* L. Burckle and

H. D. Grissino-Mayer, "Stradivari, Tree Rings, Violins, and the Maunder Minimum: A Hypothesis," *Dendrochronologia* 231 (2003): 41–45.

p. 167 *Thomas Browne wrote in 1658:* Thomas Browne, *Hydriotaphia: Urn-Burial; or, a Discourse of the Sepulchral Urns Lately Found in Norfolk* (1658), chap. 5.

p. 167 *On the windswept flanks of the White Mountains:* V. C. LaMarche Jr., "Environment in Relation to Age of Bristlecone Pines," *Ecology* 50 (1969): 53–59; www.blueplanetbiomes.org/bristlecone_pine.htm.

p. 168 *The difficulty of aging trees:* D. R. Currey, "An Ancient Bristlecone Pine Stand in Eastern Nevada," *Ecology* 46 (1965): 564–566; Charles J. Hitch, "Dendrochronology and Serendipity," *American Scientist* 70 (1982): 300–305; Darwin Lambert, *Great Basin Drama: The Story of a National Park* (Niwot, Colo.: Roberts Rinehart Publishers, 1991).

p. 172 *the stilt palm of lowland tropical forests:* J. H. Bodley and F. C. Benson, "Stilt-Root Walking by an Iriateoid Palm in the Peruvian Amazon," *Biotropica* 12 (1980): 67–71.

p. 176 *tree rings are oddly analogous to the ear bones of fish:* "NCCOS Research on Fish Otoliths Yields Key Environmental Clues," http://coastalscience.noaa.gov/news/feature/1103.html; A. Meldrum and N. M. Halden, "Fine-Scale Oscillatory Banding in Otoliths from Arctic Charr (*Salvelinus alpinus)* and Pike (*Esox lucius*)," *Materials Research Society Symposium Proceedings* 489 (1999): 167–172.

p. 178 *However, death is not the end of the line for trees:* Kevin Krajick, "Defending Deadwood," *Science* 293 (2001): 1579–1581.

p. 182 *"He that plants trees":* Thomas Fuller, *Gnomologia* (1732).

CHAPTER 7. SIGNS AND SYMBOLS

Epigraph: Marvin Bell, "Thirty-two Statements about Writing Poetry," *The Writer's Chronicle: Commemorative 2002 Edition* (Fairfax, Va.: Association of Writers and Writing Programs).

p. 188 *"where and when a party had split up":* Wade Davis, *Shadows in the Sun: Travels to Landscapes of Spirit and Desire* (Washington, D.C.: Island Press, 1998), 41.

p. 189 *the mortal maiden Daphne:* Mary Pope Osborne, *Favorite Greek Myths* (New York: Scholastic Press, 1989), 23–40.

p. 189 *"knocking on wood":* www.worldwidewords.org/qa/qa-tou1.htm; Missouri Forest Products Association: www.moforest.org/news_info/facts/slang.htm; www.pitt.edu/~dash/uppsala.html.

p. 190 *the Philadelphia Phillies:* www.carpenters.org/carpentermag/Topping910_01.pdf; *Philadelphia Business Journal,* August 12, 2003, http://philadelphia.bizjournals.com/philadelphia/stories/2003/08/11/daily16.html.

p. 190 *Its symbol is the maypole:* www.wyrdology.com/festivals/may-day/maypole.html.

p. 191 *Some associate it with Yggdrasil:* www.wizardrealm.com/norse/holidays.html.

p. 191 *Summer holidays that directly or indirectly involve trees:* www.netglimse.com/holidays/summer/midsummer_or_litha.shtml; www.odinsvolk.ca/O.V.A.%20-%20SACRED%20CALENDER.htm.

p. 192 *In fall, an important Jewish holiday, Sukkot:* www.hillel.org/jewish/holidays/sukkot/default; www.torah.org/learning/halacha-overview/chapter17.html.

p. 194 *"The cultivation of trees":* "An Address by J. Sterling Morton on Arbor Day 1885," www.arborday.org/arborday/morton1887.cfm.

p. 194 *"It is well that you should celebrate your Arbor Day":* Robert Haven Schauffler, *Arbor Day Letter of Theodore Roosevelt to School Children, Arbor Day: Its History, Observation, Spirit and Significance* (New York: Moffatt, Yard, & Co., 1913), v.

p. 195 *"Advocates of the oak":* Arbor Day Foundation, "Oak Becomes America's National Tree," December 10, 2004, http://tenfreetrees.org/media/pressreleases/pressrelease.cfm?id=95.

p. 198 *Paper money:* www.banknotes.com/lb.htm; www.banknotes.com/gh.htm; www.banknotes.com/us.htm.

p. 205 *"These scaffolds are 7 to 8 feet high": Life of Belden, the White Chief; or Twelve Years among the Wild Indians of the Plains: From the Diaries and Manuscripts of George P. Belden* (1871), ed. Gen. James S. Brisben (Athens: Ohio University Press, 1974), 87.

p. 205 *"He is gone, the lion of a man":* Alfred Burdon Ellis, *The Yoruba-Speaking Peoples of the Slave Coast of West Africa: Their Religion, Manners, Customs, Laws, Languages* (London: Chapman & Hall, 1894), 57.

p. 206 *Many institutions now plant trees:* "[Tarrant County, Texas] 2006 Crime Victim Rights Week Events," www.oag.state.tx.us/victims/2006cv_events.pdf; "National Crime Victims Rights' Week, April 23–29, 2006: Community Awareness Projects," www.ojp.usdoj.gov/ovc/ncvrw/2006caps.html.

p. 207 *"dark, hidden, near-impenetrable worth of our unconscious":* Bruno Bettelheim, *The Uses of Enchantment: The Meaning and Importance of Fairy Tales* (New York: Vintage Books, 1977), 94, 217n.

p. 208 *"Mankind, fleet of life":* Aristophanes, *Birds* (414 B.C.), in *The Complete Plays*, ed. Paul Roche (New York: Penguin Books, 2005).

p. 209 *"Their strength is secret":* Sandra Cisneros, *The House on Mango Street* (New York: Vintage Books, 1991), 74–75.

p. 210 *The word* tree *is derived from the word* dūrus*:* Winfred Philipp Lehmann, *Theoretical Bases of Indo-European Linguistics* (New York: Routledge, 1993).

p. 210 *"I never saw a discontented tree": John of the Mountains: The Unpublished Journals of John Muir*, ed. Linnie Marsh Wolfe (Madison: University of Wisconsin Press, [1938] 1979), 313.

CHAPTER 8. SPIRITUALITY AND RELIGION

Epigraph: Joseph Campbell and Bill Moyers, *The Power of Myth* (New York: Anchor Books, 1991), 258.

p. 215 *"Spirituality guides people about contentment":* The Dalai Lama, *The Good Heart: A Buddhist Perspective on the Teachings of Jesus* (Somerville, Mass.: Wisdom Publications, 1998), 7.

p. 216 *"Think of a tree":* Sogyal Rinpoche and Andrew Harvey, *The Tibetan Book of Living and Dying*, rev. ed. (San Francisco: HarperSanFrancisco, 2002), 37.

p. 218 *A "world tree" is central:* Mark O'Connell and Raje Airey, *The Complete Encyclopedia of Signs and Symbols* (London: Hermes House, 2006).

p. 218 *In the Yucatán Book of Chilam Balam:* Ralph L. Roys, trans., *The Book of Chilam Balam of Chumayel* (Norman: University of Oklahoma Press, 1967); see also http://members.shaw.ca/mjfinley/creation.html; www.arthistory.sbc.edu/sacredplaces/trees.html; www.sacredearth.com/ethnobotany/sacred/worldtree.php.

p. 218 *The ancient Egyptians believed:* Barbara S. Lesko, *The Great Goddesses of Egypt* (Berkeley: University of California Press, 1999).

p. 219 *Kabbalah envisioned a world tree:* Daniel C. Matt, *The Essential Kabbalah: Heart of Jewish Mysticism* (San Francisco: HarperSanFrancisco, 1995).

p. 219 *The Sami people:* H. R. Davidson Ellis, *Myths and Symbols of Pagan Europe: Early Scandinavian and Celtic Religions* (Syracuse, N.Y.: Syracuse University Press, 1988).

p. 220 *the symbolic Tree of Life:* Israel Regardie, *The Tree of Life: An Illustrated Study in Magic* (St. Paul, Minn.: Llewellyn Publications, 2000), 49–54;

J. H. Philpot, *The Sacred Tree: The Tree in Religion and Myth* (Felinfach, Wales: Llanerch, 1994).

p. 221 *"Why were Adam and Eve allowed":* Personal communication from John McLain.

p. 221 *"all things share the same breath":* Attributed to Chief Seattle, www.ilhawaii.net/~stony/seattle2.html.

p. 221 *described the archetypal Indian as "deeply religious":* Angie Debo, *The History of the Indians of the United States* (Norman: University of Oklahoma Press), 4.

p. 222 *The relationship between the people and the land:* Åke Hultkrantz, *Native Religions of North America: The Power of Visions and Fertility* (San Francisco: Harper & Row, 1987), 25.

p. 223 *"It is a ceremony of sacrifice and thanksgiving":* Interview with Lorain Fox Davis and Tsultrim Allione on connections between Buddhism and Native American practices, www.taramandala.org/Article_TS_LFD05.htm.

p. 224 *"In a tree's highest boughs":* Hermann Hesse, "On Trees," in *Wandering: Notes and Sketches*, trans. James Wright (New York: Farrar, Straus, and Giroux, 1972; © Frankfurt am Main: Suhrkamp, 2005), 57–58.

p. 224 *"Trees in particular were mysterious":* Carl G. Jung, *Memories, Dreams, and Reflections* (New York: Vintage Books, 1963), 67–68.

p. 226 *"When you enter a grove peopled with ancient trees":* Lucius Annaeus Seneca, *Epistolae morales ad Lucilium* 4.41 (64 C.E.).

p. 228 *The Western Ghats:* www.sacredland.org/world_sites_pages/Sacred_Groves.html; A. Anthwal, C. Ramesh, R. C. Sharma, and A. Sharma, "Sacred Groves: Traditional Way of Conserving Plant Diversity in Garhwal Himalaya, Uttaranchal," *Journal of American Science* 2 (2006): 35–43; M. Gadgil and V. D. Vartak, "Sacred Groves of India—A Plea of the Continuous Conservation," *Journal of the Bombay Natural History Society* 72 (1975): 313–320.

p. 228 *Trees and spirituality are intimately linked:* Jean Markale, *The Druids: Celtic Priests of Nature* (Rochester, Vt.: Inner Traditions International, 1999); www.aoda.org/articles/druidry.htm.

p. 230 *the Osun-Osogbo Sacred Grove of Nigeria:* http://whc.unesco.org/en/list/1118.

p. 231 *"A few minutes ago every tree was excited": Meditations of John Muir: Nature's Temple*, ed. Chris Highland (Berkeley, Calif.: Wilderness Press, 2001), 3.

p. 233 *In the holy writings . . . over twenty species of trees:* John Paterson and Katherine Paterson, *Consider the Lilies: Plants of the Bible* (New York: Crowell, 1986); Michael Zohary, *Plants of the Bible* (New York: Cambridge University Press, 1983).

p. 234 *"even when all hope is lost, planting should continue":* Mawil Y. Izzi Deen, "Islamic Environmental Ethics, Law, and Society," in *Ethics of Environment and Development: Global Challenge, International Response,* ed. J. Ronald Engel and Joan Gibb Engel (Tucson: University of Arizona Press, 1990), 194; see also http://environment.harvard.edu/religion/religion/islam/index.html.

p. 235 *"Do not be deserters":* www.islamic-world.net/khalifah/khulafa_ur_rashiduun1.htm.

p. 235 *Tu B'Shvat, the New Year of the Trees:* Stephen Butterfass, "Tu B'Shevat," www.ny054.urj.net/living/Tu_B'Shevat/TuB'Shevat.shtml; Barak Gale and Ami Goodman, "The Trees Are Davening: A Tu B'Shvat Haggadah Celebrating Our Kinship with the Trees and the Earth," www.coejl.org/tubshvat/documents/tub_haggadah.php; www.shalomctr.org; Michael Strassfeld, *The Jewish Holidays—A Guide and Commentary* (New York: Harper & Row, 1985).

p. 240 *"For me, trees have always been the most penetrating preachers":* Hesse, "On Trees," 57.

p. 241 *"What is necessary":* Rainer Maria Rilke, *Letters to a Young Poet,* trans. Stephen Mitchell (New York: Random House, 1987), "Letter Six: December 23, 1903."

CHAPTER 9. MINDFULNESS

Epigraph: Wendell Berry, "Planting Trees," *Collected Poems* (New York: North Point Press, 1985). Copyright © 1984 by Wendell Berry. Used by permission of the author.

p. 245 *"charred and limbless trunks of trees":* Cormac McCarthy, *The Road* (New York: Alfred A. Knopf, 2006), 7, 18, 76, 41, 25, 74.

p. 246 *"Nearly every morning I go to the attic":* Anne Frank, *The Diary of a Young Girl* (New York: Anchor Books, 1996), 176.

p. 247 *Anne's tree appears to be unsavable:* "Anne Frank's Chestnut Tree to Be Cut Down," *New York Times,* November 15, 2007; "Anne Frank Tree Escapes Axe Again," BBC News, November 21, 2007, http://news.bbc.co.uk/2/hi/europe/7105235.stm.

p. 248 *a thriving moss-collecting industry:* J. L. Peck and P. S. Muir, "Conservation Management of the Mixed Species Nontimber Forest Product of 'Moss'—Are They Harvesting What We Think They're Harvesting?" *Biodiversity Conservation* 16 (2007): 2031–2043; J. L. Peck and P. S. Muir, "Inventory and Regrowth Rate of Harvestable Tree and Shrub Moss in the Oregon Coast Range," *Forest Ecology and Management* (in press).

p. 248 *endangered marbled murrelet:* International Union for Conservation of Nature and Natural Resources, 2007 Red List of Threatened and Endangered Species, www.iucnredlist.org/search/details.php/2989/all.

p. 248 *Recent studies have shown that moss:* N. M. Nadkarni, "Colonization of Stripped Branch Surfaces by Epiphytes in a Lower Montane Cloud Forest, Monteverde, Costa Rica," *Biotropica* 32 (2000): 358–363; A. R. Cobb, N. M. Nadkarni, G. A. Ramsey, and A. J. Svoboda, "Recolonization of Bigleaf Maple Branches by Epiphytic Bryophytes Following Experimental Disturbance," *Canadian Journal of Botany* 79 (2001): 1–8.

p. 251 *"You will find something more in woods":* Saint Bernard of Clairvaux, Epistle 106, from *Encyclopaedia Britannica*, 9th ed. (1878), vol. 3.

p. 251 *"I went to the woods":* Henry David Thoreau, *"Walden" and "Civil Disobedience": Authoritative Texts, Background Reviews, and Essays in Criticism*, ed. Owen Thomas (New York: W. W. Norton, 1966), 61.

p. 252 *"there were on the planet where the little prince lived":* Antoine de Saint-Exupéry, *The Little Prince*, © 1943 by Harcourt, Inc. and renewed 1971 by Consuelo de Saint-Exupéry; English translation © 2000 by Richard Howard, reprinted by permission of Harcourt, Inc.

p. 252 *"If you want to build a ship":* Antoine de Saint-Exupéry, *The Wisdom of the Sands*, translated by Stuart Gilbert (Harcourt Brace and Co., 1950).

p. 253 *He declared these to be "reserved":* Quoted in Michael L. Morrison, Bruce G. Marcot, and R. William Mannan, *Wildlife-Habitat Relationships: Concepts and Applications* (Washington, D.C.: Island Press, 2006), 379.

p. 254 *"Today we are faced with a challenge":* Wangari Maathai, Nobel Lecture, Oslo, Norway, December 10, 2004, http://nobelprize.org/nobel_prizes/peace/laureates/2004/maathai-lecture.html.

p. 255 *Circle of Life Foundation:* www.circleoflifefoundation.org.

p. 256 *Gear for Good:* www.gearforgood.org.

p. 257 *Global Canopy Programme:* www.globalcanopy.org.

p. 257 *Forests Now Declaration:* www.forestsnow.org.

p. 259 *Evangelical Climate Initiative:* www.christiansandclimate.org.

p. 261 *The costs of planting this unsuitable nonnative tree:* Dov F. Sax, "Equal Diversity in Disparate Species Assemblages: A Comparison of Native and Exotic Woodlands in California," *Global Ecology and Biodiversity* 11 (2002): 49–57.

p. 261 *Although birds do frequent eucalyptus groves:* "Deadly Eucalyptus," www.prbo.org/OBSERVER/Observer108/Focus108.2.html; Ted Williams, "America's Largest Weed," www.audubonmagazine.org/incite/incite0201.html.

RECOMMENDED READINGS

GENERAL WORKS

Balog, James. *Tree: A New Vision of the American Forest.* New York: Barnes & Noble, 2004.

Harrison, Robert Pogue. *Forests: The Shadow of Civilization.* Chicago: University of Chicago Press, 1992.

Heinrich, Bernd. *The Trees in My Forest.* New York: Cliff Street Books, 1997.

Hora, Bayard, ed. *The Oxford Encyclopaedia of Trees of the World.* Oxford: Oxford University Press, 1981.

Lowman, Margaret D., and Nalini M. Nadkarni. *Forest Canopies.* San Diego, Calif.: Academic Press, 1995.

Lowman, Margaret D., and H. Bruce Rinker. *Forest Canopies.* 2d ed. San Diego, Calif.: Academic Press, 2004.

Nadkarni, Nalini M. "In the Treetops: Life in the Rainforest Canopy." In *World Book Science Year Book* 2003: 54–67.

Nadkarni, Nalini M., and Nathaniel T. Wheelwright. *Monteverde: The Ecology and Conservation of a Tropical Cloud Forest.* New York: Oxford University Press, 2000.

Pollan, Michael. *The Botany of Desire: A Plant's Eye View of the World.* New York: Random House, 2001.

Suzuki, David, and Wayne Grady. *Tree: A Life Story.* Vancouver, BC: Greystone Books, 2004.

Suzuki, David, and Amanda McConnell. *The Sacred Balance: A Visual Celebration of Our Place in Nature.* Vancouver, BC: Greystone Books, 2002.

Tudge, Colin. *The Tree: A Natural History of What Trees Are, How They Live, and Why They Matter.* New York: Crown Publishing, 2006.

Wilson, Edward O. *Biophilia.* Cambridge, Mass.: Harvard University Press, 1984.

BOOKS FOR CHILDREN

Collard, Sneed B. *In the Rainforest Canopy.* Tarrytown, N.Y.: Marshall Cavendish, 2006.

Johnson, Jinny, and Nalini Nadkarni. *Rainforest* (Kingfisher Voyages). Boston: Kingfisher, 2006.

Williams, Judith. *Exploring the Rain Forest Treetops with a Scientist.* Berkley Heights, N.J.: Enslow Publishers, 2004.

INTRODUCTION

Benzing, David H. "Vascular Epiphytism: Taxonomic Participation and Adaptive Diversity." *Annals of the Missouri Botanical Garden* 74 (1987): 183–204.

Maslow, Abraham. *Motivation and Personality.* New York: Harper, 1954.

CHAPTER 1. WHAT IS A TREE?

Cole, Rex Vicat. *The Artistic Anatomy of Trees.* New York: Dover, 1965.

———. "Oldest Living Tree Tells All." *Nature and Culture* 3 (2): 78–90.

Hallé, F., R. A. A. Oldeman, and P. B. Tomlinson. *Tropical Trees and Forests: An Architectural Analysis.* Berlin: Springer-Verlag, 1978.

Horn, Henry S. *The Adaptive Geometry of Trees.* Princeton: Princeton University Press, 1974.

Kozlowski, T. T., P. J. Kramer, and S. J. Pallardy. *The Physiological Ecology of Woody Plants.* San Diego: Academic Press, 1991.

Mattheck, Claus, and Helge Breloer. *The Body Language of Trees: A Handbook for Failure Analysis.* Berlin: Springer-Verlag, 1994.

Thomas, Peter. *Trees: Their Natural History.* Cambridge: Cambridge University Press, 2000.

Zimmermann, Martin H., and Claud L. Brown, with Melvin T. Tyree. *Trees: Structure and Function.* Berlin: Springer-Verlag, 1971.

CHAPTER 2. GOODS AND SERVICES

Cooke, Giles Buckner. *Cork and the Cork Tree.* New York: Pergamon Press, 1961.

Lewington, Anna. *Plants for People.* Leatherhead, Surrey: Bartholomew Press, 1990.

Nearing, Helen, and Scott Nearing. *The Maple Sugar Book, together with Remarks on Pioneering as a Way of Living in the Twentieth Century.* White River Junction, Vt.: Chelsea Green, 2000.

Parsons, James J. "The Cork Oak Forests and the Evolution of the Cork Industry in Southern Spain and Portugal." *Economic Geography* 38 (1960): 195–214.

Taber, George M. *To Cork or Not to Cork: Tradition, Romance, Science, and the Battle for the Wine Bottle.* New York: Scribner, 2007.

Usher, George. *A Dictionary of Plants Used by Man.* London: Constable, 1974.

CHAPTER 3. SHELTER AND PROTECTION

Coutts, M. P., and J. Grace. *Wind and Trees.* Cambridge: Cambridge University Press, 1995.

Kaplan, Stephen. "The Restorative Benefits of Nature: Toward an Integrative Framework." *Journal of Environmental Psychology* 15 (1995): 169–182.

Kitchen, James W., and William S. Hendon. "Land Values Adjacent to an Urban Neighborhood Park." *Land Economics* 46 (1967): 357–60.

Kuo, Frances E., et al. "Fertile Ground for Community: Inner-City Neighborhood Common Spaces." *American Journal of Community Psychology* 26 (1998): 823–51.

Nelson, Peter, and Judy Nelson, with David Larkin. *The Treehouse Book.* New York: Universe, 2000.

Ulrich, Roger S., and David L. Addoms. "Psychological and Recreational Benefits of a Residential Park." *Journal of Leisure Research* 13 (1981): 43–65.

CHAPTER 4. HEALTH AND HEALING

Carpman, Janet R., Myron A. Grant, and Deborah Simmons. *Design That Cares: Planning Health Facilities for Patients and Visitors.* 2d ed. New York: Jossey Bass, 2001.

Gerlach-Spriggs, Nancy, Richard Enoch Kaufman, and Sam Bass Warner Jr. *Restorative Gardens: The Healing Landscape.* New Haven: Yale University Press, 1998.

Malkin, Jain. *Hospital Interior Architecture: Creating Healing Environments for Special Patient Populations.* New York: Van Nostrand Reinhold, 1992.

Murray, Elizabeth. *Cultivating Sacred Space: Gardening for the Soul.* San Francisco: Pomegranate, 1997.

Ulrich, Roger S. "Natural versus Urban Scenes: Some Psychophysiological Effects." *Environment and Behavior* 13 (1981): 523–553.

CHAPTER 5. PLAY AND IMAGINATION

Curtis, Charles H., and W. Gibson. *The Book of Topiary.* Rutland, Vt.: Charles Tuttle, 1986.

Greenwald, Richard M., Lori H. Penna, and Joseph J. Crisco. "Differences in Batted Ball Speed with Wood and Aluminum Baseball Bats: A Batting Cage Study." *Journal of Applied Biomechanics* 17 (2001): 241–252.

Louv, Richard. *Last Child in the Woods: The Nature-Deficit Disorder.* Chapel Hill, N.C.: Algonquin Books of Chapel Hill, 2005.

Nadkarni, Nalini M., and Teri J. Matelson. "Bird Use of Epiphyte Resources in Neotropical Trees." *Condor* 91 (1989): 891–907.

Patton, P. "Wooden Bats Still Reign Supreme at the Old Ball Game." *Smithsonian* 15 (1990): 152–165.

Smith, David B. *Curling: An Illustrated History.* Edinburgh: John Donald Publishers, 1981.

CHAPTER 6. CONNECTIONS TO TIME

Lara, Antonio, and Ricardo Villalba. "A 3620-Year Temperature Record from *Fitzroya cupressoides* Tree Rings in Southern South America." *Science* 260 (1993): 1104–1106.

Pannella, Giorgio. "Fish Otoliths: Daily Growth Layers and Periodical Patterns." *Science* 173 (1971): 1124–1127.

Stephenson, Nathan L., and Athena Demetry. "Estimating Ages of Giant Sequoias." *Canadian Journal of Forest Research* 25 (1995): 223–233.

Stokes, Marvin A., and Terah L. Smiley. *An Introduction to Tree-Ring Dating.* Tucson: University of Arizona Press, 1996.

CHAPTER 7. SIGNS AND SYMBOLS

Calvino, Italo. *The Baron in the Trees.* Translated by Archibald Colquhoun. New York: Random House, 1959.

Campbell, Joseph. *The Mythic Image.* Princeton: Princeton University Press, 1974.

Cook, Roger. *The Tree of Life: Image for the Cosmos.* New York: Avon Books, 1974.

Jung, Carl G. *Man and His Symbols.* London: Aldus Books, 1964.

Merrill, Christopher, ed. *The Forgotten Language: Contemporary Poets and Nature.* Salt Lake City: Peregrine Smith Books, 1991.

CHAPTER 8. SPIRITUALITY AND RELIGION

Bernstein, Ellen, ed. *Ecology and the Jewish Spirit: Where Nature and the Sacred Meet.* Woodstock, Vt.: Jewish Lights Publishing, 1998.

Dimont, Max I. *Jews, God, and History.* New York: New American Library, 1962.

Erdoes, Richard, and Alfonso Ortiz, eds. *American Indian Myths and Legends.* New York: Pantheon Books, 1984.

Leggett, Trevor. *A First Zen Reader.* Rutland, Vt.: Charles Tuttle, 1960.

Nhat Hanh, Thich. *Living Buddha, Living Christ.* New York: Riverhead Books, 1995.

Paterson, John, and Katherine Paterson. *Consider the Lilies: Plants of the Bible.* New York: Thomas Crowell, 1986.
Walker, Winifred. *All the Plants of the Bible.* New York: Harper & Row, 1957.
Wall, Steve. *Wisdom's Daughters: Conversations with Women Elders of Native America.* New York: HarperCollins, 1993.

CHAPTER 9. MINDFULNESS

Hill, Julia. *The Legacy of Luna: The Story of a Tree, a Woman and the Struggle to Save the Redwoods.* San Francisco: HarperCollins, 2000.
Maathai, Wangari. *The Green Belt Movement: Sharing the Approach and the Experience.* New York: Lantern Books, 2003.
———. *Unbowed: A Memoir.* New York: Anchor Books, 2007.
McCarthy, Cormac. *The Road.* New York: Alfred A. Knopf, 2006.

POETRY CREDITS

John Ashbery: "Some Trees," from *Some Trees*, by John Ashbery. Copyright © 1956, 1978, 1997, 1998 by John Ashbery. Reprinted by permission of Georges Borchardt, Inc., on behalf of the author.

J. Daniel Beaudry: "Breath," 2007.

Marvin Bell: "The Self and the Mulberry," from *Nightworks: Poems 1962–2000*. Copyright © 2000 by Marvin Bell. Reprinted with permission of Copper Canyon Press, www.coppercanyonpress.org.

Wendell Berry: Excerpt from "I Go Among Trees and Sit Still," from *A Timbered Choir: The Sabbath Poems, 1979–1997*. Copyright 1999 by Wendell Berry. Reprinted by permission of the publisher. "Planting Trees," from *Collected Poems*. New York: North Point Press, 1980. Copyright 1984 by Wendell Berry. Used by permission of the author. Excerpt from "Woods." From *Collected Poems*. Copyright 1987 by Wendell Berry. Reprinted by permission of the publisher.

Sophie Cabot Black: "The Tree," copyright 2004 by Sophie Cabot Black. Reprinted from *The Descent* with the permission of Graywolf Press, Saint Paul, Minnesota.

William Blake: "Laughing Song," from *Songs of Innocence and of Experience*, 1789.

John Henry Boner: Excerpt from "The Light'ood Fire," from Edmund Clarence Stedman, ed., *An American Anthology, 1787–1900*. Boston: Houghton Mifflin, 1900; Bartleby.com, 2001.

George "Duke" Brady: "Forest Canopy Freestyle Rap." OSF Ruff Sumthin Productions, 2002.

Henry Cuyler Bunner: "The Heart of the Tree," 1884.

Moyra Caldecott: "Fern-Leafed Beech," from *The Breathless Pause: Poems and Thoughts*, by Moyra Caldecott. © Bladud Books, UK, 2007.

John Calderazzo: "Douglas Fir, Falling," copyright by the author, 2007.

John Clare: "In Hilly-wood," *The Rural Muse.* London: Whittaker, 1835.

Billy Collins: "Splitting Wood," from *Picnic, Lightning.* © 1998. All rights are controlled by the University of Pittsburgh Press, Pittsburgh, PA 15260. Used by permission of the University of Pittsburgh Press.

E. E. Cummings: Lines from "i thank You God for most this amazing." Copyright 1950, © 1978, 1991 by the Trustees for the E. E. Cummings Trust. Copyright © 1979 by George James Firmage, from *Complete Poems: 1904–1962* by E. E. Cummings, edited by George J. Firmage. Used by permission of Liveright Publishing Corporation.

Louise Erdrich: "I Was Sleeping Where the Black Oaks Move," copyright © by Louise Erdrich, reprinted with permission of The Wylie Agency.

Pancho Ernantes: "I Am a Peach Tree," translated by Robert M. Laughlin, from *The Tree Is Older Than You Are: A Bilingual Gathering of Poems and Stories from Mexico with Paintings by Mexican Artists,* selected by Naomi Shihab Nye. New York: Simon & Schuster, 1995.

Robert Frost: "The Road Not Taken."

Pam Galloway: "Arbutus, on Galiano," from *Saving Trees, Saving Wildwood: Poems for the Future.* Gabriola, BC: Reflections Publisher, 2001.

Federico García Lorca: "Romance Sonambulo," from *The Selected Poems of Federico García Lorca,* translated by Stephen Spender & J. L. Gili, copyright © 1955 by New Directions Publishing Corp. Reprinted by permission of New Directions Publishing Corp.

Michael S. Glaser: "A Blessing for the Woods," first published in the *Christian Science Monitor,* May 19, 1999. "The Presence of Trees," copyright Michael S. Glaser. Both poems reprinted by permission of the author.

Woody Guthrie: Excerpt from "Remember the Mountain Bed." *Mermaid Avenue, Volume II.* Music by Billy Bragg & Wilco. New York: Woody Guthrie Publications, Inc., 2000.

Jane Hirshfield: "Tree," from *Given Sugar, Given Salt.* New York: HarperCollins Publishers, 2001. Copyright © Jane Hirshfield. Reprinted by permission of HarperCollins.

Li-Young Lee: Excerpt from "From Blossoms," from *Rose.* Copyright © 1986 by Li-Young Lee. Reprinted with the permission of BOA Editions, Ltd., www.boaeditions.org, and Bloodaxe Books, UK, 2007.

Li Po: "Conversation among Mountains," translated by David Young, from *Five T'ang Poets: Wang Wei, Li Po, Tu Fu, Li Ho, Li Shang-Yin.* Oberlin: Oberlin College Press, 1990.

Lynn Martin: "Under the Walnut Tree," from *Blue Bowl*. Yakima: Blue Begonia Press, 2000.

Rochelle Mass: "Waiting for a Message," from *The Startled Land*. Hershey, Penn.: Wind River Press, 2003.

Gail Mazur: "Young Apple Tree, December," from *Zeppo's First Wife: New and Selected Poems*. University of Chicago Press, 2005.

W. S. Merwin: "Trees," from *Flower and Hand: Poems 1977–1983*. © 1997 by W. S. Merwin, permission of the Wylie Agency. "Place," from *The Rain in the Trees*. New York: Alfred A. Knopf, 1988.

Robert Morgan. "Translation." Copyright 2004 Robert Morgan. By permission of the author.

Vijaya Mukhopadhyay: "Wanting to Move," translated by Naomi Shihab Nye, from *This Same Sky: A Collection of Poems from around the World*. New York: Four Winds Press/Macmillan.

Pablo Neruda: "Too Many Names," from *Selected Poems* by Pablo Neruda, translated by Anthony Kerrigan, edited by Nathaniel Tarn and published by Jonathan Cape. Reprinted by permission of The Random House Group Ltd.

Ono no Komachi: "This pine tree by the rock," from *The Ink Dark Moon: Love Poems by Ono no Komachi and Izumi Shikibu*, translated by Mariko Aratani and Jane Hirshfield, copyright © 1990 by Jane Hirshfield. Used by permission of Vintage Books, a division of Random House, Inc.

Marge Piercy: "A Work of Artifice," copyright © 1970 by Marge Piercy, from *Circles on the Water* by Marge Piercy. Used by permission of Alfred A. Knopf, a division of Random House, Inc.

Kenneth Rexroth: Excerpt from "City of the Moon," by Kenneth Rexroth, from *Flower Wreath Hill*, © 1979 by Kenneth Rexroth. Reprinted by permission of New Directions Publishing Corp.

Pattiann Rogers: "The Determinations of the Scene," from *Firekeeper, Selected Poems*, revised and expanded edition. Minneapolis: Milkweed Editions, 2005.

Christina Georgina Rossetti: "Who Has Seen the Wind?" 1869.

Rumi: "Every Tree," from *The Glance: Songs of Soul-Meeting* by Jalaloddin Rumi, translated by Coleman Barks with Nevit Ergin, p. 85. New York: Penguin, 2001.

Percy Bysshe Shelley: Excerpt from "Ode to the West Wind," 1819.

Karen I. Shragg: "Think Like a Tree." Bloomington, Minnesota, 2002.

Gary Snyder: "Why Log Truck Drivers Rise Earlier Than Students of Zen," from *Turtle Island*, copyright © 1974 by Gary Snyder. Reprinted by permission of New Directions Publishing Corp.

Rabindranath Tagore: Untitled, from *Fireflies*, p. 66. Macmillan, New York, 1928.

"In Praise of Trees," from *Selected Poems,* translated by William Radice. London: Penguin, 2005. Reprinted by permission. "Recovery 6," from *Final Poems.* New York: George Braziller, 2001.

David Wagoner: "Lost," from *Traveling Light: Collected and New Poems.* Copyright 1999 by David Wagoner. Used with permission of the poet and the University of Illinois Press.

Bill Yake: "inside out," first published in *Calapooya Collage,* vol. 18, 1994. "The Tree As Verb," reprinted with permission of the author.

INDEX

Note: page numbers in italics indicate illustrations and captions.

Text: 10/15 Janson
Display: Gotham
Compositor: Integrated Composition Systems
Indexer: Thérèse Shere
Illustrator: Bill Nelson
Printer/Binder: Thomson-Shore, Inc.